THE
BLOOD
OF A YOUNG MAN

THE
BLOOD
OF A YOUNG MAN

RAYMOND A. RAMIREZ

ARPress
ILLUMINATING IDEAS
EMPOWERING VOICES

ARPress
45 Dan Road Suite 5
Canton, MA 02021

Hotline: 1(888) 821-0229
Fax: 1(508) 545-7580

Ordering Information:
Quantity sales. Special discounts are available on quantity purchases by corporations,associations, and others. For details, contact the publisher at the address above.

Printed in the United States of America.

ISBN-13:   Softcover   979-8-89330-466-4
                    eBook   979-8-89330-467-1

Library of Congress Control Number: 2024902459

Contents

DEDICATION

This book is dedicated to my mother Dolores Aguayo Ramirez Ortiz. Brothers Michael, Gilbert, Andy Ramirez, Robert Ortiz. Sister Sylvia Ortiz Castillo and two brothers now deceased Philip Ramirez and Jimmie Ortiz. And special thanks to my wife Ellen Marie Brockbank Ramirez

Chapter 1:
Introduction

A long, long time ago, there was once this young boy who lived in a garage with his mother and four brothers. The garage had no running water; they used a (Tina), metal pail, when they had to go at night. The garage was full of holes. During the winter it would get very cold. This young boy was told he was now the man of the house because his father died when his father was only 36 years old. This young man was thirteen years of age when he saw his father die. His father was a WW11 veteran. This young boy saw his father die at a VA hospital.

From that day forward he had to care for his younger brothers and mother. The boys' father left him a shine-box, it was a very nice shine-box, it was made especially for the young boy. The boy would go to school and go shining shoes at night. During the summer the boy would take the RTD, bus to downtown Los Angeles very early to shine shoes. He would shine shoes till 9 or 10 pm. He would make enough money to eat, buy clothes, would always make it a point to try to make enough to bring home $10.00 for his mother. Back then it was a lot of money.

To him, Downtown Los Angeles was a strange fascinating and exciting place to visit. It seems people from all over the world were there. People spoke different languages, looked different, and just about everyone wore shoes. It was like stepping into a portal and seeing things only a seer could see. Seers are people that can open portals anywhere. Sometimes

1

this happens by accident, people are unaware they have this capability. Sometime a person could walk out their door, look out the window, be running through a field of tall grass, or drive down a road and turn their head and see something they didn't expect. Perhaps, they would see a UFO, a ghost, a demon or maybe the past or the future. Even as recent or in the far past, seers would go in a trans and open doorways to different worlds. What you're capable of seeing is endless. There were many people throughout history that were able to interpret their visions and shared them with humanity. Here are just a few; Nostradamus, Galileo, Sir. Issac Newton, Einstein, Tesla, Steven Hawkins, Erich Von Daniken, who often asked the question "WHY". This young man would see terrible things happen at night. People being stabbed to death or shot, sometimes other older boys would try to steal his money, so he had to fight them. Sometimes he used his shine box as a weapon, that's what his father taught him.

One early morning a group of older boys walked up to him and demanded he give them his shine box, he immediately hit the biggest boy on the side of his head knocking him down to the sidewalk. The RTD bus arrived just in time to take him to downtown Los Angeles. It was quite an exciting morning already. He wondered what type of adventures he was going to have today.

On this day he was getting lots of work. He was charging .25 cents per shine. But his customers would pay him $1.00. So, he decided to watch a movie during his lunch break. He bought himself a couple of pieces of chicken, a bottle of soda; and went in to see a Japanese samurai movie. The movie was in Japanese with subtitles. The movie kept the young man on the edge of his seat. It was an extremely exciting movie.

From that day forward he was hooked for life on samurai movies. From that day, every time, a new samurai movie would come out, he made sure he was there to see it. Little did he know his uncle, who was only three years older than him, would teach him how to fight like a samurai. And he would learn different ways to fight using different Japanese fighting techniques. He would learn how to defend himself using techniques like; Karate, Aikido, Judo, and Kendo fighting. HONOR was a big part of these teachings.

This young man would always ponder on things at night when he went to bed. He would ponder about girls and things like how Levi's came about. When he couldn't find a satisfactory answer, he would go to the library and look it up. One day he asked, who, how, and why did Levi's come about. So, he walked to the library, asked the librarian for a story about Levi Strauss. For some reason, only known to himself, he didn't ask anyone for help in getting a library card. So, when he started reading the book, he found it so interesting that he decided to read the book cover to cover. When he was done reading, he had to walk at least two miles in the dark to get home, crossing over train tracks and dark streets, in a heavy gang and drug-infested area.

On his way home, he remembers living on a street named Kern. There he was growing up with his mother and father; he also had three younger brothers at the time and one other on the way. His father worked as a carpenter. When he worked, life was good. But when he would get laid off, the pendulum would swing the other way. When he first realized things were about to go very bad, was when he was sitting on the curb of the street in front of their rented house. He could remember seeing his dad driving up the street at an extremely rapid speed. When his dad drove up the driveway, the young man ran up to great him. The young man arrived just in time to see his dad running down the driveway and the undercover cop pull up behind his father's car, jumped out of his car, pull his revolver out, and shouted, "Stop Mon, or I'll shoot." Mon was his fathers' nickname. Everyone who knew his dad would call him Mon. The police knew him well. His father continued to run and as he leaped over the fence, the officer fired two shots hitting the top of the wooden fence. You could see the splinters from the wooden fence fly into the air. His father got away that day as he would on many other days. However, there were times his dad would get caught out there, they didn't catch him with any heroin on him. So, they would give him like thirty days of jail time. But this just made it harder for the family.

Chapter 2:
The Brave Young Heart

One day this young man was playing with his neighbor, and they climbed up on top of the garage next door. It wasn't a well-built garage; it was put together using scraps of wood. There were pieces of wood stacked on top of the roof. When they decided to get down, the neighbor decided to pull the wood panel that the young man was standing on. Suddenly, the young man falls through the roof and hits the ground hard. He ended up almost biting his tongue off and breaking his right arm. His mother and neighbors all came out to see what had happened because they all heard the noise of all the wood crashing down onto the ground. When his mother asked him to open his mouth, she let out a scream because she saw his tongue was in pieces and barely hanging on by a piece of skin. His right arm was broken, he was a mess. His mother ran and got a towel from inside and asked the neighbors if they can give them a ride to the hospital. This young man's father was in jail and wasn't there to help.

During this time his mother was pregnant and was due sometime in February. She was only two months away from having her fifth child. On the way to the hospital the boy was trying to console his mother. He would be rubbing her back; he would not cry because he didn't want his mother to cry and be tormented by what has just happened. When they arrived at the emergency hospital, they just rapped his tongue together and told them, he needed to go to the General Hospital because they

5

couldn't treat him there. So off they went to the trauma center at the General Hospital. This young boy and mother waited for hours before being seen. It wasn't until a doctor came walking by and noticed the boy bleeding heavily that the boy finally was taken in.

By then, the boy started getting worse, he was running a high temperature; he was drinking a lot of his blood because he didn't want to let his mother see all the blood coming out. He also lost almost all of his baby teeth. He was six years of age at the time. He was a very brave young man. The doctors put him in a tub full of ice and covered him in it. They had to bring his fever down and the swelling before they could operate on him. All of this was very agonizing for the young man. When he went into surgery, he could see all the lights above him, with doctors and nurses all around the table. They had to sew his tongue together without any anesthesia. But he hung in there; he didn't want to disappoint his mother.

He was in the hospital during Christmas. He received a gift from his mother, teacher, the doctors, and the nursing staff. His teacher came to visit him while hospitalized and gave him a little clear plastic Christmas tree with a lot of tiny blue balls with a few red and green ones. He became acquainted with hospital life; the nurses would help him with his writing skills and a little bit of math. Before long, he was on his way home. He didn't return to his school; which was Reagan Elementary School. After the boy arrived at home his father was released from jail. His father decided he was going to buy a house for them all. So, he wanted to move and start fresh.

Below is a picture of the young man, his brothers, and his neighborhood friend; who pulled the board away from him, causing the young man to fall.

Just before they were ready to move the boys' father was arrested again. Things got pretty bad quickly. There was no food in the house, his mother was unsure as to what to do. Therefore, she decided to send the boy to her mother's house.

Chapter 3:
Precious Memories in Sugar Shack

The Sugar Shack

The boys' mother wrote a note and put him on the bus. So, the boy went on to his grandma's house for the first time on a bus and by himself. Everything he saw out the window of the bus was new and strange to him. However, the young man was brave, although he was scared, he didn't want to show it. Suddenly, they were on a street he recognized. So, when they arrived at the corner of Hammel Street, he stood up and told the driver he has arrived and to please let him off the bus. The bus driver opened the doors of the bus and like lightning; he jumped off the bus and ran to grandma's house.

Grandma's house was affectionately called the sugar shack because everyone would get together and play music on their guitars, maracas, and other musical instruments to create beautiful sounding music. They would all laugh, dance, play, and eat there. Yes, it would become a happy, joyful, experience like no other. Most of the time there would be nothing but boy cousins there. There were a lot of testosterones around. At times the young bucks would test their antlers but would cease and all would be well. There was never any disrespect in the house; neither on the grounds nor to the neighbors. It was like one big family; neighbors would look out for each other's children. One rarely sees that nowadays.

On this day, however, one of his beautiful aunts was there visiting. She read the note he brought to the house. She didn't have any money or transportation, so, they walked back to the young man's house. When they got to the house a decision was made to gather everyone including his pregnant mother and the whole tribe walked to grandma's house. By the time they arrived, he was very tired, thirsty, and hungry.

Shortly thereafter, another beautiful aunt appeared with her tribe. The family was starting to grow. Then the grandmother gave the young man some change and asked the boy to go and buy some corn tortillas, beans, and rice. Grandma already had a large sack of potatoes. Somehow, grandma could cook for an army. She would whip up delicious, healthy meals for everyone. For some reason, unknown to the young man, everyone was experiencing financial difficulties. His grandfather and grandmother was taking care of everyone who showed up. But the most wonderful thing happened at the Sugar Shack, there were singing and playing musical instruments every day of the week through every season.

The young man's family ended up moving about a mile away. To a street named Blanchard by Gage Street. From there the young man had to walk to Hammel Street Elementary School. This was about a mile or two away. To get to school he had to walk over a dirt hill, walk through allies so he could walk the shortest route. When it rained things just got a little dicey and crazy. Most of the time, they couldn't afford shoes for all of them. So they would take milk cartons and cut out soles for their shoes. Because the milk cartons had a wax covering and it made them waterproof, it worked well.

This young man used to like to make things; he was great at figuring out how to make things. He made his first zip gun when he was in the third grade from no blueprints or guide. He remembers visiting his grandparents' house on Hammel St, in East Los Angeles. Now this was a uniquely special place, it was an enchanted house. It was called the Sugar Shack. This was a home like no other, both his grandparents, two uncles, and two cousins lived there. His two uncles would play the guitar and sing beautiful Mexican songs. Their picture is shown below.

The Young Man's Grandparents

Brothers and Uncles

They could have sold records. Times were very hard back then, his uncles were laid off from work, so all the families would take their children down to the grandparents' house to eat. He had no idea as to how his grandmother did it. But with a sack of potatoes, a big bag of pinto beans, a huge bag of rice, a big bag of masa, a red box of lard, lettuce, and tomatoes, she managed to feed everyone.

The summer nights were precious and magical. The sugar shack was located on the far backside of the yard. The only light came from the two windows in the front of the house and the porch light. The lighting wasn't bright at all, so when the cousins looked at the dark sky above, they were able to see the twinkling stars. The young mans' grandfather was the patriarch, and his grandmother was the matriarch of the family.

The young boy's father was a good caring and loving father. He could play the guitar, sing, and loved to dance especially fast dancing. He had lots of friends. Later in life, his mother would tell him stories about his father. He remembers his mother telling him that she would catch his father dancing with a bunch of girls at the nearby corner of the block. His goal was to buy a house for his family. His father worked as a carpenter. His father helped build many of the buildings still standing in Long Beach, California. His mother once told him, his father was once working as a "Pan-man" and fell off a beam. His father fell from a high beam and landed on a floor pan below. He was injured and out of work for several months. His father was a lover of life. He once hit a pigeon while driving home. When he arrived home the pigeon was still stuck to the grill of the car. It was dead. He asked the young mans' mother to prepare the pigeon for his dinner, so the bird would not have died in vain.

Once the family was driving down the road when his father came to a complete stop. After a minute or so, the boy's mother asked, "why are we stopped". His father replied, "There is a family of ants crossing the road. I'm letting them cross to the other side." He also remembers when they all got out of hand and their mother would shout "enough is enough!" She would tell their dad to hit them for misbehaving and put them to bed. They all ran to their room and got on their beds. They were all scared and frighten, as their dad walked in holding a long, large belt across both hands. He snapped the belt, and they all heard the snapping sound; they started to whimper and shiver as he approached. Suddenly he raised his index finger to his lips, and a whisper came out from his lips.He said, "I'm not going to hit you, but when I hit the bed, sound like I'm hitting you." They all finally fell asleep.

Chapter 4:
Bittersweet life

The Young Man's Father

The boy remembers his father taking them to the drive-in, just about any day of the week. His father would first stop at a hamburger place on Third and Gage Street and buy either five hamburgers or five hot-dogs for a dollar. He would buy enough for them all to eat at the drive-in. The boy remembers seeing movies like; Pinocchio, Snow White, and Mr. Bojangles, These movies were very astounding, incredibly moving, tearjerkers. But, well made movies, he thought Disney was on a roll. He was amazed at his mother's ability to know everything that was going on. It was like she had eyes in the back of her head. Oh yes, she was so good at knowing when they were up to no good, she would put a stop to it in a heartbeat.

One winter evening, it was cold, raining and very dark, a stranger came and knocked on their front door. The stranger called out for their dad. "Mon" he shouted, "Mon" he shouted once again. Their father started towards the door. Their mother stops him and warns him that this is not a good idea. She goes on and tells him, "Don't go outside." Their father is well aware of her capabilities, so he heeds her request. Seconds later everyone could hear the air coming out from the tires of their car. It was a hit! Their father was very fortunate that evening. Yes, their father was doing great selling heroin. He was making lots of money. Their father took their mother to a dealership the very next day and bought a new car outright with cash. The young boy remembers his father pulling out a huge wad of money rapped in a rubber band. The young man thought, oh we're rich. But that wasn't the case. He was getting ready to buy a huge amount of heroin to make even more money. Since he was doing so well his parents were putting money away for their dream house. Their mother decided to sew the extra money into the sofa. They had stacks and stacks of money hidden there.

One evening, it was still light outside, his parents decided to go to the supermarket and buy groceries and goodies for a movie night at home. They started gathering the kids together, but the young man said, "I'll stay home and take care of things." His mother asked, "are you sure you want to do this?" The young man replied "yes." They all drove away, leaving the young boy behind. He was having a grand old time. He was changing the channels to the TV, watching what ever he wanted.

Soon it started to get dark, everyone was still gone. He was beginning to worry. The sky was getting darker and darker. Then it became totally dark. He started to get scared. He slowly crawled, on his hands and knees, to the kitchen. He pulled out a knife and crawled back to the living room next to the sofa, and sat on the floor with his back to the wall. There he waited for his family to return. The young man was scared to death. When he looked out the window, he could see eyes looking back at him. He looked at the doorways and could see images or shadows of people coming close. When his family did return, he heard the car pull into the driveway. He jumped with joy, ran to the kitchen to put the knife back and smiled as he greeted everyone back home. He was a big boy now.

Their father gathered everyone one day and said, "Let's go for a ride." "We are all going to ride in the new car," his father said. They were in an area unfamiliar to the boys. The homes were all big and beautiful looking. Everywhere they looked the lawns were nice and green; the trees and plants were beautiful looking. As their dad was about two blocks from his destination, an undercover roadblock sprang into action. Their father had nowhere to go. An undercover police officer quickly ran up to the car window and said, "Mon, you are under arrest. If you do what we ask, things may go better for you". Their dad agreed. He was asked to attend his meeting, act normal, and pretend he is unaware of the raid. He was also told the entire area was surrounded and under tight surveillance. Their dad proceeded to his meeting. As he was approaching his meeting place, he turned to the boys and said, "listen up, quickly stand up and block the rear window" the boys did as requested. To the young man, this was like opening another doorway, a portal to get to the other side. "What other side, the young man asked himself, the entire area is surrounded. There were undercover officers crawling everywhere." There were undercover officers posing as gardeners, up on telephone poles, looking down on everything going on. They were even parked four or five houses down in an unmarked car.

Things looked very grim for his dad but as they approached the house where the meeting was to take place. The boys had the rear car windows blocked; their dad swapped seating positions with their mother. Their dad asked their mother to slow down almost to a stop when they arrive in front of the house. Then she was to step on the gas and speed away. Their dad jumped out of their moving car. The race was on. Their mother was doing her best to get away; she was driving fast, turning right, and turning left with undercover officers in close pursuit. Unfamiliar with the area, their mom turned into a dead-end street. The family was caught. The undercover officers really laid into their mother. She was even threatened with jail time. They told her they were going to take the kids away. It was a vulgar seen. But the officers drove away looking for their dad. Amazingly, their dad vanished in thin air. It was like he stepped into another dimension.

(Dolores Aguayo Ramirez Ortiz)
The Young Man's Mother

Chapter 5:
Mischief choice

The young man thought to himself; Houdini doesn't have anything on my dad. His grandmother, on his father's side, once told him, "When one is born, he or she is placed on a journey of life." She explained that everyone is given choices on this journey. Depending on the choices one makes will determine the end results of your life. Doorways will open and shut. Be careful on opening a doorway you choose to travel on because thunder only happens when it rains. You may experience a quantum entanglement because time is a constant. You could stop a clock, but you could never stop time. Time is like a river; you can't step in the same river twice. You may be able to bend time and space, but you cannot stop it. Like our universe, it continues to expand. Why? The young man's father was like time itself. No one could stop his motions. His expansion, he was his own universe. We just happened to live in it, the young man theorizes.

His father needed to buy another huge amount of heroin to continue his expansion, to purchase their new home. One day, off they all go in their new car. They arrived in a neighborhood very similar to the one they were in with the police. This time they noticed signs reading (restricted area), the young man asked, "what does that mean?" His mother replied, "I don't know." She just didn't want to explain what it meant to the young man. They arrived at their destination; his father parked the car, got out and walked toward the house. He knocked on the door, walked in, and disappeared from their sight. One hour later he walks back to the car, and they all drive off.

On the way home, his father becomes increasingly paranoid. He decided to ask the young man for his help. His father had bundle of balloons filled with heroin capsules. These balloons are rubbers filled with heroin capsules.His father gave him two huge balloons to swallow; if the police showed up. The young man's mother went ballistic. She stated, "Absolutely not," they were arguing for blocks regarding his father's decision. His father stated, "Look, the rest will be my responsibility. I'll need to eat them. He is a big boy, he can handle it'". The young man became scared but at the same time didn't want to disappoint his dad.

Sitting in the back seat of their car, the young man started to pray and pray. He prayed that there would be no police officers showing up while they were on their way home. They finally arrived home safely. The boy was so relieved they had no interruptions, on their journey home. Very early the very next morning, around one or two; everyone in the house awoke to a loud smashing noise at their front door. The family could hear men yelling as they rushed into their home. The young man saw his father run past his doorway wearing nothing but his underwear and juggling bundles of heroin. His father ran into the bathroom, locked the door, and began to flush the bundles of heroin down the toilet. As the young man watches undercover officers pass his doorway, the young man contemplates jumping off his top bunk, onto rushing officers. But calculates the distance to be too great. He could hear the officers trying to knock open the bathroom door. His father not only had the bathroom door locked, but he also had his full body leaning against the door. You could hear the pounding and yelling until his father flushed his last bundles of heroin and started squeezing out the small rear bathroom window. As the undercover officers punched through the door, the first one, stuck his hands into the toilet to grab as many balloons of heroin as possible.

By then, it was like a portal opened and swallowed his dad. In an instant, wearing nothing but his underwear, he vanished. The young man wanted his father to get away, but thought, this is impossible, how could this be. But yes, his father got away that day, only to run again another day. While his dad was on the run, he thought he had managed to flush all the evidence down the toilet. However, that was not the case, the undercover officer, who put his hands down the toilet, managed to pull enough bundles of heroin to send him to prison for a very long time. His father was finally caught while hiding out at a friend's house. He had many friends. He was a very likable person. He would give his shirt off his back, if needed too.

When his father learned he could be sentenced to prison for a very long time; he decided to ask the boy's mother, to get him the very best lawyer in town. His mother did as he requested. The lawyer promised everything, but all the lawyer did was take all their money. The boy remembers his mother saying to him, "Your father is going to prison," on this day and at this time. She allowed the young man to go see him for the very last time. He was also allowed to take two of his younger brothers with him under very strict conditions. He was to keep his brothers safe, and not to allow their father to see them. On that day, the three young men went to see their father taken away by a black and white bus. This was an adventure for him. As the young men walked their way to the Third Street Park, where their father was being held, the young man thought to himself, what doors he might open on this journey to see their dad. When the three brothers arrived at the park the young man found a grassy knoll with a shade tree. "This is where we will hide he said," they had a good view, they were hiding in the shade, and the wait began. Suddenly a black and white bus appears.

Around one hour later, men in chains and wearing jump suits started coming out of the building. They could see sheriff's standing by with high powered rifles. They could see the men in chains and jumpsuits as they began to the bus. Suddenly, all three young men could see their dad. His hands were cuffed; his feet were tied with chains. They could see the men load the bus in pairs. None of the young men made a noise. They just stared in sadness; their dad was now gone. They waited there until the black and white bus was gone.

Chapter 6:
Life in the Garage

On their way home the young man asked himself, "Why and then, how will we survive". Things became extremely rough for them all. The family was in crises. One very hot summer day, the young man went and laid on his bed. He closed his eyes and was able to see a place that was very green. There were lots of trees, rivers, and sand all over the place. He could also see mountains in a distance. He continues, looking at this unfamiliar place. He noticed the Sun was extremely bright. He could see strange creatures moving around the ground, like snakes, that are usually associated with life and death. There was an unfamiliar scent in the air. It became so intensely hot, it was almost became unbearable. But he continued, he moved closer to the mountains. Suddenly a loud crack broke the silence, thunder sounded nearby. Then a deluge of rain started to pour as though the sky had opened up; as if he was in the presents of Tho. The rain kept pouring down. The young man thought to himself, this is not a place I wanted to be. He struggled to return to his bed. He started to see strings of green and red lights moving towards each other. The red would stand out so intensely on the ground, on leaves, bushes, and on himself. It was like the devil himself was rushing through there. He knew he had to get away quickly or forever get caught in this river. He started to frantically back pedal, flapping his hands, arms, like a hummingbird, with the exception of also pushing with his legs and feet. The young man was determined not to give up.

Just then he saw this strange hummingbird, like creature, reach out for him. It was his mother shaking him to wake up. The young man woke up drenched in sweat. He immediately knew this was not a door he wanted to reopen. He was fortunate he made it back that day, for time stops for no one. Did he find a way to bend time? He thinks to himself and states, "I'm going to be just like you dad." He soon starts a hobby of raising pigeons. He raised tumbler pigeons; they were lots of fun to watch fly high up into the blue sky and tumble down a way. Everyone loved to watch his flock of pigeons fly. One day, one of his beautiful black and white tumblers started tumbling down and down until it hit the ground. To the young man, the pigeon was caught in a perpetual state of motion. The pigeon had entered into a time continuum of no return. The stage was set, and the primordial gravitational pull was greater than the pigeon could handle. Like a small comet, coming in flaming hot, his pigeon hit earth, leaving an empty space in the young man's heart.

One day his mother took them to a nearby church carnival. It was crazy fun; they all came home with a fluffy yellow baby chick. The boys had this plastic blue one foot swimming pool. To their unbelievable surprise, and horror, their youngest brother thought the baby chicks could swim like ducklings and put all five of the chicks in the swimming pool. Their mother sensing something was wrong, quickly ran to the pool in time to see all the chicks at the bottom of the pool. She quickly pulled them out of the water and placed them on her apron. She immediately ran inside the house, placed them on the kitchen table and commenced to compress their chest. Water was oozing from their beaks. She valiantly tried to save them. There was one chick that couldn't raise its head. She placed them all on a towel in the open oven door. Two of the chicks moved close to the chick that was non-responsive, the two chicks moved side by side to the unresponsive chick and spread one of their wings over the chick. To their surprise and hopes, soon after, the chick revived. Unfortunately all five chicks were eaten by a cat the very next day. Now, all the chicks were gone.

The next day the young man walked into the garage only to see feathers all over the garage. The cat had paid his pigeons a visit. The young man couldn't comprehend all the bad things happening in his life. The young man found a wooden box and quickly turned it into a trap to capture the cat and take it far away from his pigeons. Every time the boy went into the garage to feed his pigeons, he would check the trap. But days went by then weeks went by and no cat. His pigeons were being left alone too.

One day his mother asked him, "How would you like to move in with your grandmother," the one living on Lovett Street. She goes on to say, "I need you to check things out for all of us, because we are all going to move there too." The young man was eager to start a new adventure, he was ready to step into another dimension, another doorway, what will he see, he was so ready for this. The young man packs a few things to take with him. He tells his brothers to take care of his pigeons, but he forgets to tell them of the trap he laid out for the cat. They all climbed into the car and off they go to grandmas' house; his father's parents. The young man was welcomed with open arms. His mother and brothers all visited, ate, had fun, and left him there with his grandparents. He was soon enrolled into Eastman Elementary School, about eight to ten blocks away. He soon made friends with his next-door neighbor who also went to the same school. His new friend was a year older and was at the next higher grade level. These two young men soon became the best of friends. They would walk to and from school each school day. On weekends, they would hang out and play together. The young man's grandparents were raising him as their own. They were also raising his cousin, a beautiful young girl that was about three years younger than him. She was the daughter of his dad's younger brother; who was also spending time in a state penitentiary. Her mother gave her up when she was just a baby. They were growing up as though they were brother and sister. They were being taken care of very well. The young man's grandmother was very religious. Every Sunday they would walk to church. For some reason, unknown to the young man, he never saw his grandfather go to church. He soon had his own rosary and learned to pray with the rest of the flock. This went on for about a year or so. The young man's grandmother was very knowledgeable in herbal medicine which had been passed on to her by their ancestors. When the boy would get a severe earache, tooth ache, or stomach problems his grandmother would ask the boy to pull off leaves from certain trees and plants growing in their back yard. These herbs worked extremely well.

After about a year or two his mother and brothers moved in. When this happened the young man moved into the garage with the rest of his family. His mother made the garage as comfortable as possible for all of them. They had a gas stove but no bathroom or running water. This became home, it was very neat and kept clean. But life there had its ups and downs. It all depended on what doors you wanted to walk through. The garage had electricity, so they had lights and a television to watch. They had lots of fun there. The young man started up his hobby and his love for pigeons again. He built a pigeon cage on the corner of the yard, next to the garage. Soon he had tumblers again and would let them fly high in the sky. Now he could watch them, once again, tumble in the sky against the white clouds. His best friend soon received permission to raise pigeons too. The young man was fascinated by electricity and magnetism. He would scrape his magnet on across the dirt and pull up what looked like splinters of metal., He would place the slivers of metal on a white piece of paper, place the magnet underneath and move the metal slivers around to form a variety of shapes.

One day while his pigeons were flying, he noticed a sparrow hawk flying and making a circular flying pattern, just hovering in the sky. He then noticed it had a baby sparrow clutched to its talons, but the weight of the baby sparrow and gravity itself was having an adverse effect on the sparrow hawks flight pattern. The weight and gravity were bringing it down. At first glance it looked as though the sparrow hawk was coming down right on his head. But a flap of the sparrow hawk's wings gave it a last-ditch effort to stay afloat. The young hawk started to glide and descend to the front of the yard. From the back of the yard, the young man commenced to run to save the baby sparrow. The young man, the sparrow hawk, clutching the baby sparrow, all arrived across the street at the same time. The sparrow hawk started to fly away by releasing its prey and taking flight. he young man jumped high into the air and grabbed the sparrow hawk in midair. That's when he noticed the baby sparrow was missing its head. Boy, "'Wow, what an incredible catch" he said to himself. The sparrow hawk was magnificently, and strikingly beautiful. Nature has provided this magnificent bird with all the colors in the rainbow. It was a very young sparrow hawk that was probably out honing its survival skills. The young man already had an empty cannery bird cage. Which is where he decided to put his sparrow hawk. The young man fed his hawk raw hamburger meat and water. He wanted to train the hawk to become a hunting bird. When his grandfather learned of his catch; the grandfather asked the young man to show it to him. The grandfather looked at the sparrow hawk and asked, "What are you feeding it?" He replied, "Small pieces of hamburger meat." His grandfather was wearing a glove and stuck his hand in the cage and grabbed the young hawk. He turned to his grandson and said, "This is a very dangerous bird, it could take your eyes out." The grandfather moved over towards the large slab of black tar pavement floor in front of the garage, raised his arm that was holding onto the bird and slammed it on the black slab pavement floor. His hawk made a loud screaming noise. The bird was dead. The young man said to himself, "Why didn't he just turn it loose". The young man didn't know that his brush, with this sparrow hawk, was a glimpse into a portal that would appear to him in the future. It would be a doorway that would forever alter his mind. But life in the garage was good. They were always having lots of fun, and everything was an adventure. The young man and his best friend would end up fighting each other. It was always a brawl, like two gladiators fighting for survival.

Chapter 7:
Courage and Bravery

One day they started a rock fight. Soon other boys joined in on his best friend's side. The young man's four brothers came out to join the rock battle. His best friend's team had an empty lot full of rocks to use as ammunition. They had overwhelming continuous rocks for fire power. Then a small window on his mothers' car was broken. The young man became furious. He picked up the same rock, walked over to the side of the best friend's garage and threw it through the side garage window, shattering it to pieces. The young man seeing this was dangerous for all of them, decided to attack. He dashed to the front of the yard, jumped over their chain-link-fence, ran over to the empty lot. He spotted his best friend and charged towards him. He caught his friend by surprise. His best friend gets up to get away, but it was too late. Like a tiger grabbing a gazelle, it was just too late. The young man grabbed him by a veracious headlock with his left arm and commenced to hit his head and face with his right hand. His best friend gets away and runs to the back door of his house. The young man leaps and once again grabs his friend with a headlock and begins to slam his head on everything around him, like the corner of his friend's mother's sawing machine table. His best friend breaks away once again, running into the long kitchen and begins to pull out kitchen knives and throwing them at him. The young man breaks away and walks home, and walks back into his home, the garage.

A few days later they became best of friends again. One day his best friend told him, He wished that he could live in the garage with the young man and all his brothers. His best friend was an only child. When the young man and his best friend were attending Stevenson Junior High School, things in the neighborhood began to change dramatically. Some gang members had moved in and wanted to control the neighborhood. Around the same time a new family moved in the two-story house across the street. There were two beautiful girls and their brother along with both their parents. It was around this time the young man gave up playing cowboys and Indians. He became aware that girls were meant for hugging and kissing. He realized all that he had been missing.

Yes, now the young man spent an awful long time in his grandparents' bathroom. He was showering every day, making sure he was squeaky clean. He would brush his teeth day and night. His teeth were so bright; they would gleam with every smile each and every night But he had no game. He didn't know how to approach girls. Living with boys and growing up with nothing but boys, he just didn't know how to act around girls. So, he asked his mom, "Mom, how do I meet girls, what should I say to them?" His mother replied, "When you see her just wave and say hi." "That's it!" says the young man. Now he sits on the street curb, just waiting for one of the girls to come out to play. He waits and waits, but they just don't come out.

One day, he finally sees both the girls playing in the front yard. However, he couldn't get their attention. This frustration goes on and on. Week's go by and finally he sees them out in the front yard watering the grass and plants and occasionally wetting each other with the hose. He could see they were barefoot, wearing shorts and a t-shirt. They just looked so pretty to him. He decided to cross the street and enter their world. As he gets in front of their hedge; he stops and waves to them and says hello. To his surprise, both girls responded in kind. They started asking him a lot of questions, like; how's the neighborhood, the schools, the stores and a lot more. The girl's brother came out to call the girls back inside the house; they needed to do some chores inside. All three became good friends. It was the summer of 1960. The young man was only eleven years old; his Cowboy and Indian days were now over. He had just set foot into another realm, another dimension. It was a doorway from which he would never be able to return. He continues to take steps down this journey not taking small steps but full bold steps. Now, he wanted to see what was on the other side. Sometimes things would jump out at him as to test his resolve. But the young man would continue to push forward. When the brother of the two girls asked the young man if he could join him on his shoe shinning business, the young man replied "Yes!" in a resounding voice.

For now, he had someone to help fight the daily "Dogs" of Downtown. The area where the young man lived was very nice. All the homes were of good size. The lots were big, there were trees all over the neighborhood and close to the street curb in front of every house was a beautiful fig tree. People could just reached up and grabbed one. They were very delicious. But the neighborhood gang was getting bigger and more troublesome. After the gang failed to recruit the young man and his best friend, every time they would get spotted, the gang members chased them trying to catch them to beat them up. But the two best friends were very strong, healthy, they didn't drink nor smoked. Both played sports, they were well disciplined not only by their coaches but by their parents or on the young man's case, "parent". This gang just roamed around the young man's universe. From time to time their worlds would collide. The two best friends would bolt, like comets in the skies. They would just streak through the neighborhood. Jumping over fences, climbing onto roof tops, leaping from one house to the next, walking across blocked walls, and presto they would vanish. The two best friends would do this almost daily. Every time they needed to go to the store for their parent, one would walk over to each other's house and whistle a secret whistle, and both would go off to the neighborhood store to get what was needed.

One day a gang member who lived down on the next street, from the young man, was out riding his bicycle on the sidewalk. He was headed straight towards the young man. It was like two knights jousting; right in front of the young man's grandparents' house. The two continued to get closer and closer; both wondering who was going to flinch first. At the very last micro second the boy on the bike hit his breaks, as the young man stepped to the side slightly. Then the young man pushed the gang member and his bicycle into the chain-link fence. The gang member was trapped. For some reason, the young man didn't strike him, but gave him a few choice words. Upon release, the gang member sped away quickly saying a few choice words of his own. The young man walked into his yard and went into the garage, where he lived. About one hour later the entire gang was out front calling the young man out. The young man was unaware of what was going on out front because the garage was way back of the yard. His beautiful cousin came running to the garage and said the entire gang was out front calling for him. She also went on and said, "My dad wants you to come into the house." Her dad was the young man's uncle, the younger brother of his dad. His uncle was now out of prison, a free man after spending almost all of his youth in prison.

When the young man walked into his grandparents' house, his uncle was sitting in the living room. The uncle proceeded to give the young man some advice. He says, "Look these guys will be after you forever, it's time to put a stop to this right now. Go out there and ask for the biggest and baddest guy they have. I know you have the capability to beat any one of them. I have seen you fight before, you're a very good fighter." The young man was very scared; his knees were shaking and knocking. However, not wanting to show his fear he boldly walks up to the screen door, opens it, and sees a sea of gang members. They were everywhere on the street, and the sidewalk. The young man walks towards the front driveway gate looks around and wonders, what door was he about to open. Will they all jump me as I step out, he thinks to himself. However, he learned even gang members have a code of honor.

The young man steps onto the sidewalk and makes a remark to the gang member who started all this in the first place. The young man acknowledges him by calling him a few choice words. Then he turned to all the gang members and stated, "I want to fight the biggest and baddest gang member". All the gang members started laughing and falling to the ground because they were laughing so hard. Soon this big huge guy comes out from the bunch of gang members pushing them aside, boldly walking towards the young man. As this gang member approaches him, the young man prepares himself for battle. As soon as the gang member gets within reach; the young man strikes like a rattle snake with a kick to the knee cap. The gang members buckle over in pain, slowly retreating his position. The young man leapt forward with quick left and right jabs to the gang member's face. This huge gang member didn't know what hit him. The young man continued his flurries of kicking and punching this huge guy. A gang member seeing that his friend was losing the fight, quickly threw him this big wavy knife. The young man quickly takes his belt off. He began to hit the gang member every time he moved close to him. The young man would struck him with the buckle side of the belt. This put a tremendous amount of pain on his opponent. All of a sudden the young man's grandmother comes running out swinging a belt around her head; chasing everyone away. The gladiator show was over; the grandmother proudly walked her grandson home. After that fight, the gang members left the young man and his best friend alone.

Chapter 8:
Advice That Stays Forever

His best friend's parents had bought the empty lot right next door to them. They were having a new house built and the contractor had a porta-potty placed out on the ease-way of the sidewalk. During the night about five gang members climbed inside the porta-potty to sniff glue. This was a way for them to get high. So, the young man's best friend saw them all climb into the porta-potty and shut the door on themselves. He went over and tells the young man what he saw. They fomulated a plan to tip the porta-potty over while the gang members were still inside. They agree to the plan and both move out to conduct their assault on the gang members inside the porta-potty. They quietly move in position to conduct their assault. Using hand signals they create a countdown of three. Using their fingers they commenced to count. With all their might they pushed the porta-potty over. You could hear screams come from inside the porta-potty; because the gang members universe just came crashing down onto the ground. The gang members were no longer flying high, they were now grounded. As the gang members tried to open the door to the porta-potty; the two young men ran just a few houses down and began to laugh at the downed gang members. It was a deliriously hilarious, spectacle. The gang members came out totally confused as to what had just happened to them.

Life continued, seasons would pass. Christmas season was here. Everyone was trying their best to make life festive and bright. They placed red, green and blue lights on the front of the garage. Everyone was merry. But at night during the winter season it would get very cold inside the garage. They sometimes had to use newspaper to plug up some of the larger holes in the garage. Their mother would turn on all the stoves burners and oven to make the garage warm. Sometimes they had to sleep with their jackets on. One day to all their surprise their dad came home. He looked very nice. He was wearing a suite and a nice looking hat. He was finally sent home. Unknown to all of his children, he was released on a humanitarian basis. His father was dying of cancer. Everyone was happy "Mon" was home. The young man just turned thirteen years of age and he was attending Stevenson Junior High School. At the time the young man was struggling with English, history, and spelling, but was excelling in electronics, science, and math. The young man really loved electronics. His teacher was teaching them how to make radios, telegraph keys, and they were taught Morris Code. His teacher had a very unique, special way of teaching his students. As all his students in class were building their own telegraph key, he had them learn each letter of the alphabet in Morris Code. When their telegraph keys were completed; the class would be seated in their chairs and the teacher would start keying out words, with a twist. The teacher started making jokes about each student. If you didn't learn the code you wouldn't know if he was making jokes about you. This really became a lot of fun. The things he would say about his students were crazy hilarious. He showed them how to make a radio without any electricity, no batteries, by just having to connect to the ground, making an antenna. The use of a tiny crystal to power the radio. Then by using some small earphones you had this forever lasting radio to tune in to many channels too. Everyone who saw it work was amazed by it. This young man built himself an electric chair. He borrowed the transformer from his brother's train set. He liked it because it had a lever to control the amount of output power. His dad saw this and wanted to try it. His dad sat in the chair, grabbed the handles, which would transfer the power to the subject. His dad liked it. He asked the young man to shock him every day.

Each day his father would ask for higher and higher voltage. Until one day the transformer blew. That was it for the electric chair.

His dad started losing his health rapidly. His father was now bed ridden. One day, a large black beautiful butterfly came into the garage and landed at the foot of his bed. His father starred at it and said, "It is time, they are coming for me." His father just starred at the butterfly as it slowly but boldly flapped it's wings still standing at the foot of his bed. It looked as though they were both looking at each other. Then their dad asked his kids to open the screen door and allow the butterfly to pass and fly off. They did as their dad requested. The butterfly flew off to continue with its business.

That weekend the young man was invited to a party; this would be his first real party. He had a girlfriend that lived up the block from him. He invited her. On Friday they both show up together to the party. The girl, doing all the inviting, was young beautiful looking. She appeared to be living alone, or maybe her parents were just gone for the weekend. There was good music playing, people were dancing and having a good time. As the night progressed, the girl having the party walked over to the young man and sat on his lap. She just didn't care about his girlfriend. The girl began to kiss his ears, his lips, and started rubbing him all over. The young man's girlfriend walked up to him and said we should leave. This was the very first time any girl has gone this far with him. The young man tells himself, "this flower has opened up, and he was very close to getting the nectar." He just couldn't pass this up; this is what every thirteen year old young man strives for. The young man tells his girlfriend to go ahead, go home; that he would be leaving soon. His girlfriend gave the young man an ultimatum. "If I leave without you, I'll never be your girlfriend again." He looks at her and says, "bye." She was gone. He continued to party with his new found girl. Soon, people started to leave and go home. They were all parting like crickets in the summer time, not worried about the harsh fall. It was getting late. But, the sounds went on. She grabbed him, took him into the bedroom and continued to hug and kiss him. The young man respected her wishes. But he was just a puppet on a string. She was in full control. It soon became about two a.m. the next day. The young man finally realized his hope of having some nectar wasjust a dream a flitting wish. It was late and he had to get home. When he arrived at the garage, the door was locked. He said to himself, "Oh no, how am I going to get in." He decided to move to a side window of the garage, he manages to push the bottom half of the window up to open it. He didn't want to wake anyone up. He slid through the window and into the garage. As he turns to shut the window, his father turned on a lamp. He called him over and gave him a tongue lashing for staying out so late. His father did give him some advice. He said, "son, don't ever disrespect a woman. Women are for hugging, kissing, and loving. Take care of them and always treat them right, and they will take good care of you. If not, they can leave you living your life in misery." This advice will stay with the young man forever, no matter what doors he opens. Within a week, his father turns gravely ill. On August 14, 1962, he dies in the V. A. Hospital in West Los Angeles. The young man is thirteen-years-of-age. The young man was happy and sad, sad that his father had now passed, and happy his father was no longer suffering in pain. The young man new he had to step out of this space-time-continuum for this dimension had now closed.

Chapter 9:
Purple People Eater

The young man made a giant leap into a new portal, where he gets involved in sports, attends parties from time to time and makes many friends. He takes a newspaper delivery job, but has no bicycle, and he continues to shine shoes. On his sixteenth birthday, his mother gave him her Cadillac. He rushed over to the DMV with his mother to obtain his drivers permit. He passed with flying colors. He had been driving since he was only twelve-years-of-age. Now, he and his friends went to work on the Cadillac, they had cut the front leaf springs to lower the front end, they also painted the Cadillac purple using paint brushes. They ran out of paint halfway through the paint job. They also obtained very nice shinny chromed hubcaps using a five-finger discount.

The Cadillac was truly something out of ELA, and also out of this world. When you saw the Cadillac coming or going, it was like something coming out of the clouds. It was like something coming out of another universe. From either side, it looked like two cars humping down the road. But, this was his car. At sixteen, he and his friends were very adventurist and daring now. One of his friends came up with an idea of driving to Arizona. His friend went on to say he had a cousin living in Clifton, Arizona. They decided to leave on a Friday evening. They choose the evening so they could cross the desert during the night. It would be much cooler at that time. It was summer and everyone was on summer vacation. They all agreed to chip in for gas, but meals were on each individual. They were all going in his good friend's car. It was a very nice looking Pontiac. The doorway opened, it was time to jump through and see what was on the other side. All four jumped into the car and drove to a gas station to fill it up. At that time gas was less than .25 cents per gallon. With a full tank of gas, off they drove into the night. When they reached an unfamiliar area; something hit the underside of the car. The car started to lose fuel rapidly. Here they were on their own, in the dark universe, with no way to get help. The driver asked, "Should I continue driving until we find a gas station or stop now and try to fix the problem before we run out of gas." Everyone voted to stop and fix the problem, as .25 per gallon was expensive to them.

It was like a carload of Mayan engineers; they were going to fix the problem. The young men didn't know it at the time, but the Mayan engineering was already embedded in their DNA. They pulled to the side of the freeway. In total darkness, with no lights or flashlights these Mayan engineers prepared to fix the car. The driver and owner of the car crawled under the car and began to feel for the leak. After several trys he located the leak. The driver determines that the hole on the gas tank was about twice the size of a BB. Three of them started to look around for anything that was the proper size to plug the hole. After an exhausted search, of finding tree twigs, someone found a pencil. It was a broken pencil but it looked as though it would work. The lead of the pencil was gone but this left a perfect cone shape to fit the hole. The pencil was placed and pushed in enough to take hold, and then tapped in with a rock. It fit perfectly; the gas swelled the wood of the pencil making a tight fit. They did it, off they drove to Arizona.

Soon, questions arose, "Will the car make it all the way to Arizona?" After quick consideration, they decided the car was unsafe and unreliable to make it to Arizona. They got off the freeway, drove to the other side of the freeway and back to East Los Angeles they traveled. They were all very disappointed. But they were only fifty miles out of Los Angeles; they really had a long way to go. During their trip back home, someone came up with an idea. He asked the young man, "Why don't we take your car?" The young man replied. "I don't know if it could make it. We've never driven it that far."

Once again, they all put it to a vote. The decision was, "Let's do it." they all shouted. When they arrived at the young man's home, they quickly unloaded the car. Then they placed a thirty gallon can of fresh water and some snacks into the young man's car. Off they went to park his friend's car in his own yard. Now they went to a gas station filled up and got right onto the freeway once again. When the Cadillac got on the freeway, it was like a spaceship pushing through the clouds. The purple and blue Cadillac was comfortable and drove smoothly down the freeway; as if a giant magnet was pulling it forward. It hummed, it was like a bomb. It was now very late, but they continued to push on. The young man was driving and the Cadillac was handling great. They were listening to music, beating on the thirty gallon water can like if it was a set of drums or bongos. They were just singing their way to Arizona. Daylight showed up and they needed to stop for gas. The young man was tired of driving, so his friend took the wheel. It was a big relief for the young man. Now he could kick back, relax, and really enjoy the scenery.

Finally, they arrived at their destination, Clifton, Arizona. The Cadillac swooped down on the small town of Clifton, looking as though an alien space craft had just descended into town with four young men aboard. They decided to drive around to see how big the town was and possibly find interesting things in town. Now, the town was very small and everyone seemed to be inside their homes. Everyone in town seemed to be afraid to come out, possibly fearing an Alien abduction. The bomb they were driving in would have scared anyone who first saw it. This purple people eater (Cadillac), drove and handled beautifully. They had no problems what's-so-ever. All four young men settled decided to go get something to eat. They all walked to a nearby hamburger joint to get something to eat for dinner. The cousin's parents did offer all of them dinner, but they did not want to burden the family financially. When they were done eating, it was good and dark.

On their way back to the house, bats were flying all around them; some even came close to their heads. But they all pushed on, showing no fear.Upon arriving at the house, they all decided to sleep out on the porch. It was nice long porch, filled with a sofa and plenty of chairs to sleep on. They were also enjoying the night and the Clifton, Arizona sky with all its brilliant stars above. The very next day they awoke early, wanted to clean up and get on with their adventure. This was a portal where all four young men stepped through. It was like bending time itself. One day, they were in East Los Angeles, and the very next day they were in Clifton, Arizona. Now they have found a river in which they could go swimming. The water was very cold and moving along. They were there swimming for hours. Then one of the boys spotted the head of a snake poking out the rocks just a foot above the water line. Once they positively identified it as a snake, they all ran out of the water. Now reality set in, the snake had taken over their swimming hole. All four of them started throwing rocks at it to scare it off and away. The young man, who owned the purple people eater (Cadillac), threw a rock the size of a fifty cent piece hitting the snake right on the head, knocking it out. The snake just went limp and was hanging out of the rock formation. The young man picked up a small tree branch shaped as a Y and proceeded to cautiously back into the water. The rest of the boys followed him. Upon arrival at the snake, the young man begins to slide his tree branch under and using the tree branch to pull the snake away from the rocks. He manages to get fifty percent of the snake onto his tree branch, balancing it on the V part of the branch. They were all walking and keeping an eye out on the snake. The young man tripped on a rock below, and the snake began to slide off the tree branch. When the snake went back into the water, the most amazing thing happened.

All four young men leaped out of the water and started kicking their feet so fast they appeared to be running on top of the water. Someone watching them would have said, "They are truly not from this earth." Now that the snake was loose in the water, they decided it was not safe to go back into the water. They went to the house and sat around the cousin's nice big porch. They were enjoying themselves eating snacks and drinking juice and having great conversations. Near the end of day the cousin came up with an activity for the next day. The cousin asked, "How would you guys like to go shooting tomorrow." They all responded with a resounding, "Yes!" Their conversation changed to weapons. One of them asked, "What type of weapons do you have?" He replied "I have three thirty odd six's and two twenty-two rifles with plenty of ammo for all. Although none of the young men had fired a weapon before, they were excited to do it.

Early the next morning the cousin's mother made everyone, eggs, bacon, toast, and orange juice. Now they were ready to go shooting. The cousin decided to use his car and drive them around, to see the land. Plus he was familiar with the area. They arrived at a wide open area that is normally used to shot weapons by the locals. The cousin informed the young men that there were wild rabbits out there; so they were excited to hunt for rabbits. They didn't want to shot targets; they wanted the thrill of the hunt. After hours and hours of hunting for rabbits not one single rabbit was seen. Someone came up with an idea, that was to split up and form two hunting teams that will cover a wider area. It sounded good, but in reality, it was a bad idea. After about a half hour a small bird landed on a tree branch, the tree was leafless nothing but dead branches were hanging on that tree. Unknown to each other, they were directly opposite of each other. Both teams spotted the bird at the same time. Without hesitation or the ability to see each other, both teams opened-up to shoot the bird down. To their surprise and shock they were in a firefight. They started to yell at the top of their lungs, Stop, Stop, don't shoot we're right here.

All the shooting stopped, not one bullet hit the bird. They would have been better off throwing rocks. Amazingly, no one was hit by the flying bullets. Fed up, they started to shot at anything mounds of dirt, rocks, at anything they desired. They soon came across a large beehive hanging from a branch of the tree. One of the young men wanted to throw a rock at the beehive;but everyone said, No, No!The young man threw his rock back down to the ground. Suddenly, the young man himself picked his rifle up and aimed straight at the beehive and started cranking off rounds. Even though, they were all telling him No, No, No, he fired anyway. But no bees were coming out. Someone said, "all the bees are gone." Another said, "let's check it for honey." Now that sounded good to all of them all. The young man walks up to the beehive and hits it with the muzzle of his rifle. Well, the hive didn't fall, but a dark cloud started to form around the beehive. They all looked at each other and said, "Run, Run!" They ran so fast that the bees couldn't catch them. There was a roadrunner running alongside them, not even the roadrunner was going to beat them to the car. They all jumped into the car and rolled the windows up. They started the car and drove safely home to the cousin's house.

That evening, they all went to the nearby hamburger joint. They started to count their money, plan the rest of their stay and the long trip home. Come to find out, they each only took ten dollars. They were all beginning to go broke. They decided to stay the rest of the week and leave on Friday to arrive home on Saturday. At that time people could buy hamburger, fries and a soda for about $1.25 or $1.50 but money was hard to come by. Their week came to an end, now it was time to drive back to ELA. That Friday morning they all got ready to leave. The cousin wanted to help out. He had a fifty gallon drum of gasoline in his yard. He provided them with a full tank of gas. He had some extra oil, so, an oil change was done.

Chapter 10:
First taste of Prejudice

Everyone hugged each other, the young men thanked his parents for their hospitality and off they went back to East Los Angeles. They all wanted to ensure they had enough money for gas and snacks for their trip home. So they put all their money together. Off they go in the purple Cadillac. As the car was climbing a mountain it started to lose power, they started getting white knuckles from the sheer worry of the car breaking down. They were worrying about not making it back home as they planned. But, once they started to go downhill the car started functioning properly. They somehow arrived at the young man's house sooner than they planned. It was like they managed to bend time itself. They made it home in one day. During their last fifty miles the Cadillac did limp home. It was dying. When they pulled up in front of the young man's grandparents' home and parked, the car died. It wouldn't start to put it in the drive way. The Cadillac died, but they all arrived home safe and sound. The young man's mother was elated to see them all home safe. His mother and the other adults didn't think they were not going to make it. They thought they would just turn around and come home. But no, the young men were determined to make it. They even sent home postcards informing their parents that they made it. To all the young men the trip was a huge success.

Now, the young man started seeing this beautiful girl that would come visit the two sisters who lived up Lovett Street. This girl was a knock out. She was petite, with jet black hair, and beautiful large green eyes.

What a doll. She was everything he wanted. He started visiting the two sisters up the block; just to see if he could see this girl more often. Maybe, just maybe, he would increase his chance of asking her out somewhere. By this time, the young man's mother had met a man. He worked in a nearby store; in the fruit department. This man had really fallen for his mother. He did everything he could to take her out and make her happy. When he finally proposed to her; he made a colossal commitment, to take care of all five of us. They both decided to save some money so they could move out of East Los Angeles. The neighborhood was starting to change. Things were starting to get very dicey. The young man's mother wanted her children to be safe and not to get involved with either gangs or drugs. But now the young man didn't want to move away. He had just found his true love. However, she was still a work in progress.

MAKING A BIG SPLASH!

Soon the young man's mother and boyfriend got married. They were ready to move out of East Los Angeles. They found a three bedroom house to rent in the city of Norwalk about twenty miles east of where they were living. Their step father was a good and hardworking man. He left the grocery store and started working with someone he made good friends with. He was now in the furniture business. Although he learned different aspects of the upholstery business; he had a deep passion in being one of the best at tying the springs on the furniture. He was fast and very good at it. He did manage to become one of the best in the furniture business. He had a deep passion for baseball, he just loved to watch the Dodgers play ball. He was also a lover of music. He would love to hear good Mexican music any time of the day or night. The young man's mother was also a lover of music; she would love to dance to the music. They were still having a hard time getting by financially. Raising five boys and paying all the bills was tough for their stepfather. The young man's mother was the Chief Financial Officer for the family so she would make sure the money stretched out until next payday.

On weekends, the stepfather would loan the young man his white Cadillac allowing him to visit with his longtime friends. But the young man had one thing on his mind; it was to go visit with his new girlfriend. She was smart, pretty, and great conversationalist. They were having great times together. He was getting along with her parents. One day he helped install a new toilet for them. He said to himself, "Wow, this is the girl I want to marry." The young man had already completed half of the eleventh grade while he was attending Garfield High School. There he excelled in Electronic, Science, and Math. The other subjects were not all that interesting to him. When he enrolled in Excelsior High School, due to his age, he was placed in the eleventh grade rather than the twelfth grade.He was disappointed he had to start the eleventh grade all over again. The school looked as though it was populated by at least eighty percent white; the rest was Hispanics, a few Orientals and doesn't recall if there were any blacks on campus, at least not in his school year.

He does recall his first taste of prejudice. It was in his geometry class, the teacher there would always get on his case, by trying to embarrass him in front of the class. The young man's teacher would always give him a zero on every paper he turned in. The teacher would call on him repeatedly to answer mathematical equations, just trying to embarrass him in front of all his classmates. One day the teacher informed the class that he was going to give a test in a weeks' time. He also advised everyone to study hard because this upcoming test was the midterm test, and a big part of their grade. Well, the young man took it to heart and studied hard every night. The midterm test finally came, everyone turned in their paper. The very next day the teacher handed out the grade results to each student. He praised those who received ninety seven percent to ninety nine percent. But then he held up one paper, and said this person got one hundred percent. Everyone was looking at each other wondering who this could be. Then the teacher called out the name of the young man. The young man was elated, smiled and proud his hard work had paid off. But this was short lived; the teacher called him a cheater in front of the entire class.

The teacher was making a big thing about this. The teacher wrote this big long geometry problem on the black board for him to solve. The young man replied, "I cannot solve that problem". The teacher said, "I'm calling the vise principle and ask him to remove you from my class for cheating". The vise principle came up and talked with both the teacher and the young man about the cheating situation. When asked by the vise principle about the cheating the young man replied, "No, I did not, sir". The vise principle decided to pull the young man from his geometry class and walked him over to the third period gym class. Now, the young man had two gym classes each day now. His six period classes were wrestling, and he was very good at that. His first semester in wrestling took him to compete in the CIF, (California Interscholastic Federation), competition, where he almost took out the first-seat for his weight division. One other gym teacher would use him to wrestle bad, tough, guys in his gym period class. The young man himself will tell you, it's better to be trained than untrained. He would just wrestle those boys to submission. The gym teacher loved that. After completing the eleventh grade, it was now summer. He was going toEast Los Angeles as often as he could to see his girl.

Chapter 11:
Becoming a Marine

One day during the month of August 1965, he was lying on the couch watching TV, when he sees a news flash, "Marines Land in Vietnam." He tells himself, "I just have to get into this war; it will probably be over in ten days." He thought. He rushes to the nearest Marine Corps Recruiting office. Upon arrival the young man tells the recruiter he wants to enlist to go to Vietnam. The recruiter asked him for his age, the young man replied, "seventeen." The recruiter tells him, "you're too young son; you'll have to get your parents' permission." The young man rushes back home with documents the recruiter gave him for his parents to sign. The young man gets his mother to sign and runs back to the recruiter and hands them to him. The recruiter tells him, "Come back this Friday that's when you and a few others will take a test if you pass, will set you up for your physical." The young man tells himself, "Gee, the war is going to be winding down by then,"

On Friday morning, the day of the test, he was at the recruiter's office at 8 am. Upon completion of the test the recruiter tells him he passed his test. Gives him a folder and tells him don't lose it others you'll have to take your test all over again. The recruiter goes on and tells him not to take anything with him when he goes to take his physical, no wallet, no watch, no keys, nothing at all. The best thing is to have someone take you there, otherwise take a bus, taxi, just don't drive yourself. If you pass your physical, they will take you for training that day. The young man tells himself, "training, no one said anything about training", and then he says, "Well, it's better to be trained than untrained". The young man figures this is best for all, He knew his high school year was going to be expensive and didn't want to be a financial burden to his family.

The day of his physical came up, one of his friends who had gone with him to Arizona, agreed to take him to downtown Los Angeles for his physical. The young man did take a few dollars with him in just in case he didn't pass his physical. He would have enough to take a bus ride back home. But that was not the case. He passed his physical and was now on his way to the Marine Corps Recruit Depot, San Diego, California. They had a greyhound bus full of future Marines headed to the depot. They were all given their own (jacket), folder to take with them and told to remember their service number before they reached the depot. As they were going south on the five freeway, the big greyhound bus passed by his home city. He tells himself, he just opened a new doorway to an unfamiliar realm. When they reached the greyhound bus terminal in San Diego, they all started to disembark and walk into a twenty-four-hour restaurant. When the young man stepped off the bus, two white men in suits grabbed him and pinned him up against the restaurant wall. Fellow recruits yelled out stating, "Hey, he's with us." The immigration agents turned him loose. The young man walked into the busy restaurant with the rest of the recruits. There he told himself, "they picked me out because of the color of my skin, I wasn't doing anything wrong." When they all were in the restaurant someone told them that a Marine Corps bus was coming to pick them up and take them all to the Marine Corps Depot.

They waited and waited, it was getting late, but he couldn't know what time it was because he left his watch at home. Finally two marine corps buses show up to pick them all up. The buses were what the marines call cattle cars. No windows, just running boards and benches to sit on. There was some yelling going on, but it wasn't bad. It wasn't all that bad, however, there were some recruits who smiled at what was being said to them. They got yelled at even more then the rest. They all went to the recruit depot. They all had to ride with their palms on their laps and starring at each other's eyes. Now, this made it worse, because now they had to stare at each other's scared dumb looks. This made some of them want to laugh. Well, they all finally arrived at the Marine Recruit Depot when they started getting off the bus the Drill Instructors (DI's) started yelling. They sounded very mean. They yelled to get in line and to put their feet on the yellow foot prints. From this day forward the young man's life was going to be, hell until graduation day.

Everything you see in the movie "Full Metal Jacket," starring R. Lee Ermey came to light. Things were just as tough or tougher. The young man was fortunate that he did a lot of pull ups, pushups, and sit ups, while in high school. Otherwise, things would have been much worse for him. He felt bad for the guys that couldn't do the exercises. The D.I.'s really had it in for those guys. The young man said to himself one day, "Where did they find these guys, because they couldn't be human. These were the toughest, meanest, people had ever encountered in his life." However, they were very good at what they were doing; because tthe unit 47 ended up taking every ribbon the Marine Corps had to offer in boot camp. They won each ribbon twice. They graduated with honors, they were Platoon 270, Regimental Honor Platoon. Out of one hundred men, eighty graduated after eight weeks of boot camp. The young man qualified as an expert at the rifle range. And, because of his test results he was assigned as an 0141, Admin Man, a clerk. This was not why he enlisted. He was upset; he wanted to be a rifleman, he was an expert rifleman; he thought surely that would send him to Vietnam. He recalls right after firing expert, the D.I. told him the smoking lamp is lit. The young man replied, "Sir, I don't smoke sir". The DI. Yells out "what, I said the smoking lap is lit, private." The DI asked for a cigarette from someone, sticks it in his mouth and lit it for him. Upon graduation they all had to report to ITR, (infantry training regimen). There they became familiar with the different weapons they may be asked to use. They also crawled underneath barbwire while live rounds from a machine gun were fired. There the young man was learning tactics.

One day the companies were having competition. The young man chose to run, he loved to run. They were going to run two miles across the foot of old smoky, a mountain everyone feared to run up it. But all west coast marines eventually do. There he learned precious tactics that will stay with him forever. While on their run, about mid-way they started to see one of their opposing companies. The young man's platoon leader asked one of the team members to pretend he couldn't run anymore. He also asked two other men to pretend they were having to hold him up as the other company passed them by, while they were marching not running to allow the straggling marine to catch up. Once the other platoon passed, they started to run again. Their platoon leader kept them at even pace. The men wanted to run faster, but the platoon leader refused. When they were returning back from where they started; before they were even spotted, their platoon leader stopped them. He put them all in proper formation, and asked everyone to sing the Marine Corps hymn as they marched across the finish line in formation. Everyone watching cheered as they proudly marched across the finish line. Unfortunately they lost the race by one tenth of a second. They all knew they could have won that race, but winning wasn't the main idea. Anyone can win a race, but it was about how they all ran the race that really counted.

After completing ITR, the young man went on ten days leave. He went home and each night he asked his step dad for the use of his car. When his ten day leave was completed he reported to Del Mar Marine Corps training base. There he had eight weeks of Administrative Training to learn how to run and function as an administrative person. He was able to go home on the weekends with the stipulation that he report back no later than ten p.m. Sunday night. Getting home was not a problem for him, but getting getting back to base was difficult.

Chapter 12:
Experience in the South

In his Admin Class he made friends; one who lived in the city next to his. One Friday evening, his friend and the friend's father brought him home. They told the young man they would be by at six pm. When that Sunday evening came about, he jumped in back of the pickup truck, they were driving. He was asked, "Hey, don't you want to sit up front." He's reply was, "No I'll be all right," off they drive to Del Mar. He was wearing dress pants, dress shows, and a dress shirt. About halfway to Del Mar, he was freezing in the back of the pickup. He laid down on the bed of the pickup to prevent the wind from hitting him directly, but he just continued to shiver. He finally started kicking and kicking the cab of the pickup truck, he couldn't stand it anymore. Finally, they decided to pull off the freeway. They didn't pull off the freeway because he was kicking the cab; they pulled off because they wanted a cup of coffee. They offered him a cup but he turned it down because he didn't drink coffee. When they got back on the road to Camp Del Mar, all three jumped into the cab and road to base, nice and cozy up front in the pickup truck; during this nice winter night. Finally he graduated tenth in his class of fifty Admin Clerks. He received orders to go on ten days leave and then to report to Camp Lejeune, North Carolina, on December 12, 1965. He was still only seventeen years of age. He spent every night taking his girl somewhere. When his leave was up, all he was basically broke. He had ten dollars to take with him to North Carolina. But he had no regrets. When his leave was up, the young man's Mother and Step Father along with a few of his younger brothers drove him to the Los Angeles Air Port to see him off. At that time, it was the longest flight he had in his life. About mid-flight a stewardess asked him if he wanted a meal. He was hungry but didn't know if he would be charged for this meal. He thought to himself, "She must be trying to get me to buy an expensive airplane meal." However, that was not the case. She was only trying to help the young marine, who was dressed in his uniform.

He arrived at Dulles International Airport wearing his winter green uniform. There was snow as far as the eye could see. It was very cold. Now he had to locate a shuttle to get him to a smaller airport where he would catch a flight to Jacksonville, North Carolina. Once there he had to catch a prop plane to get him to his destination. Taking that prop plane was the worst flight he had ever taken. The plane was going up and down as it flew tree top tall. He did notice there were trees as far as the eyes could see. There were lakes everywhere he looked. But all this just made him feel ill; he said to himself, "If this plane doesn't land soon, I'm going to be sick." The plane soon landed, the young man was very thirsty and hungry. He spots a restaurant nearby and goes in to see what he could afford to buy himself to eat. As he walked through the doors he notices nothing but older white men in there that all turned and looked at him. He thought he had walked into a dimension into the past, like if he just walked into an old western saloon. The ambiance of a western saloon was so thick you could cut it with a knife.

He continues in and sits at the bar. He ordered southern fried chicken. It was the best chicken he had ever tasted. They also gave him a bowl of water and a small towel. Never having eaten in a restaurant before, he began to drink the water from the small bowl. The man attending the bar stops him from drinking from the attendant asked "Would you want a glass of water?" Embarrassed but gladly replies "Yes." He finishes his meal and asked, "Where can I catch a ride into Camp Lejeune." The bar tender tells him where he can catch a bus into the town Jacksonville. When the young man got up to leave; someone stood up and said,

"Hell, I'll take him, I'm going that way anyway." They both leave the restaurant together. The young man throws his sea bag into the back of the pickup truck and off they go to Jacksonville, North Carolina. It was only a few miles down yonder, according to the driver. When the young man arrived at Jacksonville, he learned he had to wait two hours before the bus into the base would arrive. Just before the bus arrived he spotted a chili dog stand and ordered one. Just then his bus arrived. Now he didn't know if he should forget his chili dog or jump on the bus. Fortunately, he was able to get his chili dog. He ate it on the way to the base. He was very pleased with what he has eaten thus far. He tells himself, "Wow this southern cooking sure taste great."

He arrives at the base command center to report for duty, the command center checked him in and asked him to report to his permanent place of duty. He does so, he was assigned to 8-inch howitzer battery, which could deliver a nuclear warhead. He remembers his mom telling him there had been men in suits walking around the neighborhood asking questions about him. He told himself, "Now everything makes sense," because of the artillery battery's capability almost everyone there had to have some type of security clearance. The young man does great there; he does so good he gets promoted twice while stationed there. Even though there was a lot of racism there, he manages to make friends easily with everyone. When he first arrived there he had trouble understanding people. The majority of people there spoke with a southern draw; they said words and phrases he never heard before. He remembers many of the sleeping cubicles had confederate flags flying in them. He remembers when one of those southern boys borrowed forty dollars from him, then he refused to pay the young man back, told him so right to his face too. Well, that didn't sit well with the young man, because he may have grown up poor, but where he's from a man had his word and his honor.

One day they met each other out in town, they both ended up in the same bar. After they had a few beers together, the young man asked his friend to pay him back. He replied, no, that was a long time ago. The young man tells him let's step outside and talk more about your decision, so they both go outside the bar and walked into an ally. The young man tells him since you're not going to pay up; I'm going to have to take forty dollars' worth from your face. The fight begins, as a brawl. However, soon the young man knocks the southern boy to the ground and jumps on him and begins to pummel his face. The young man was so focused on teaching him a lesson, he didn't hear the police car pulling up behind him. He also didn't hear the police officer order him to stop. The police officer had to pull him off his friend. The police officer counseled them to shake hands and asked them to go back into the bar and buy each other a beer. The young man found it strange to be sitting across from and enjoying a beer, the guy, he had just pummeled pretty good. But they were both Marines, they were brothers, they belonged to a brotherhood only a few will ever experience. They both continued to be good friends and would run into each other in the future. Young Marines are pushed beyond their abilities to become the best they could be. Marines know this; that's why they will always be Marines. There are many brave people in this world, but then a few become Marines. But the young man was still intent to go to Vietnam. He was promoted twice in less than one year; if he stayed he probably would have continued getting promoted. He would possibly be able to retire from a great job with probably the highest enlistment man's rank available. He knew the job like the back of his hand. No, nothing mattered; he was dead set to go see what the Vietnam War was all about.

Chapter 13:
Chasing Dreams

In 1966 he starts submitting request for transfer to WESTPAC and goes on and serves in Vietnam as a combatant. He continues to submit this transfer request until he is finally given the green light. In September 1966, he is on a plan to Camp Pendleton, California. He is happy. He first goes home for ten days leave; he ends up spending those days with his beautiful green eyed girlfriend. Upon reporting to Camp Pendleton, he is assigned to the transient barracks. By now he had developed additional military skills. Due to of his easy mannerism in making friends, he was taught by his friends from the 8-inch howitzer battery how to read topographical maps, find himself on a map, and how-to call-in rounds when needed. This gave him a tremendous amount of pride in being a more knowledgeable Marine. He walked, talked, and breath Marine Corps. He was ready to go to war. But at Camp Pendleton they liked his work, so they kept him there processing outgoing troops.

The young man began to submit requests for transfer to Vietnam. Those in charge kept him in a holding pattern, until he decided to talk to the officer-in-charge. He was told they were doing him a favor, he didn't have to go to Vietnam, he was close to home. He could get promoted to corporal, if he continued working there. The young man tells the young second lieutenant, thank you. I appreciate all this but I'm determined to get to Vietnam, Sir. The butterball second lieutenant tells him, be careful of what you wish for. Within weeks of having this conversation, with the lieutenant, he receives orders for overseas duty. He is delighted and excited; he is finally going to Vietnam. He tells himself, by the time I get to Vietnam the war will be over. But he is happy he is finally on his way. They boarded a private airline jet the winter of 1966 for overseas duty. After a long flight, they stop in Hawaii. The soldiers were excited; they thought they were going to be able to paint the town. But no, as they all disembarked the plane; they were escorted into a large cage. This made them feel like animals at a zoo. Civilians were all gawking, rubbernecking and standing within five to eight feet from them. The civilians were snapping pictures of all them. The young men became irritated and mad. If those people were to look at those pictures now, they would see that about half of them never made it back home.

The young man was excited, a portal or doorway was about to open for him. He was a very curious young man, he felt like one with nature. Nature had a tremendous amount of energy; one just had to listen to the sounds of nature and become one. The plane finally lands on Okinawa an Island not too far from Japan. It was April 1, 1945, when Marines forces landed on Okinawa to defeat a well-entrenched fanatical Imperial army. As the Marines fought their way in some vicious fighting, many Japanese civilians were so scared to be captured by the Marines, they committed suicide. Women and children, whole families would jump off cliffs, because they believed the LIES, the misinformation their leader gave them. The young man tells himself, I'm glad I live in America because we are all smart enough not to believe lies where we would be committing suicide. The young Marines were taken to a transition area on a Marine base on Okinawa. There they were to prepare themselves to go to Vietnam. The idea was to build a battalion of men to send to Vietnam. A few weeks went by and now the transient area was now at battalion strength.

One early morning the battalion was called out for formation. They were all told to gather all their belongings and be prepared to be shipped out to Vietnam. At the end of his announcement the commander states, I'm calling out some names, when you hear your name step forward. He calls out three names; one of the names belong to the young man. He goes on and tells each one where to report. The young man was told to report to Headquarters Company, First Battalion 9th Marines. The young man's heart sank, this cannot be, he says to himself. I joined the Marines to go to Vietnam, why am I being held back. He settles in and begins his administrative duties. No matter where he is assigned, he always manages to do an excellent job. He pours his heart and soul into his work. He doesn't want to disappoint. He enjoyed doing his job, he had confidence in what he did,because he had good instructors. Soon he was out in town meeting the locals. He felt as though he was back home. The young man meets a beautiful Japanese girl; she is kind, respectful and took good care of him. When he visited her, she would take his shoes off, massage his feet, neck, and back. She would love him like no other. Sometimes on weekends, she would bath him. She had a big metal pan she would put out in the street in front of her home. She would prepare the water, have him strip naked in front of the house. She would use a towel to block anyone's view as he was getting in or out of the metal tub. She would softly sing as she scrubbed his body, she was incredible.

Within a few months, he was promoted to corporal. Having been promoted, had a hard time finding a way to request a transfer. But his desire to go to Vietnam was very strong. He again submitted a request to go to Vietnam. His request was again denied, but he continued to submit them. He finally is transfered to another Marine base camp where they have a transitional area. He settles down there and awaits a flight to Vietnam. Soon the squad- bays begin to fill up with Marine personal. The young man would call everyone out for formation each morning. He assigned work details, resolved issues, and exercised the soldiers. At that point in time, he was the senior enlisted man in transition unit at Camp Schwab, Okinawa. The camp was located in a unique and beautiful looking area; from their barracks you could see the beach area and the Pacific Ocean. Soon there were three sergeants added to the transition unit, the young man immediately made them platoon leaders. Now he gave the sergeants their order. He held them responsible to get the job done. The day came when everyone received their orders to Vietnam. They were told to pack their gear and be prepared to be taken to Kadena AFB at 0600 They were also told the chow-hall will be opened at 0500. Everyone in the transition unit left that day, except the young man. He was left alone with the barracks all to himself. He was soon asked to report to the first sergeant. When he reported to the first sergeant; there were four other marines ready to welcome him as the newest member of the transition team. He was told if he stayed he would be promoted to sergeant and he wouldn't have to go to Vietnam. His desire to go to Vietnam was so strong; no one was going to prevent him from getting there. He was told he would leave on the next deployment. He was disappointed he didn't get to leave with all the troops, but he was now relieved to know he was finallying leaving for Vietnam. He now had to start over and wait for another unit to be formed.

When the time came, he was determined to go with this new group, no matter what. But this was not the case, for some reason this new group was leaving first and he was leaving the following day. When the young man reported to the flight line he was asked to board a private commercial airline. During the flight he noticed that there were only a few military personnel and a lot of civilians on this flight. For a moment, he thought he was placed on the wrong flight, because it looked like it was going back to the states.He began to wonder what door he had opened and stepped through; because since joining the service he had not traveled on a single government transport. He began to tell himself something was not right, the fight continued on. He was too embarrassed to ask anyone because whatever happened, it was too late. They were not going to turn around or let him off in mid-flight. The young man fell asleep on the plane traveling to Vietnam. He began to dream of his dad; when he passed away. He recalls when the VA doctor asked him to leave the room and go out into the hallway after his father passed. The doctor wanted to speak to his mother in private. After standing out in the hallway, he became tired of standing so he went to find himself a chair. He was in building 500, of the West Los Angeles, VA Medical Center. He walked down the hallway with no chair to be found. He made a left and proceeded down another hallway and finally found a chair. While sitting there he pondered all that had occurred and began to worry about his mother and younger brothers. However, what was so unbelievable at this moment was he felt the need to go care for his mother, his chair began to move backwards. He becomes stunned and scared and breaks off his concentration as to what has just happened. He gathered himself together and tried to mentally move the chair again. Suddenly, he found himself right next to his father's hospital room. He was scared but amazed at what he was now capable of doing. He didn't want to tell anyone because he figured no one would believe what had just happened.

Little did he know that in three years he would be serving in the military himself, just like his dad? He tells himself, "I want to be just like you dad, going to be just like you." As the young man continued on this flight to Vietnam a more bazaar dream came to him. He dreams he ran into this strange woman in the basement floor of a four tier parking structure.It was nighttime, very few lights were on, it was cold and very dark; yet he could see the figure of a woman who looked to be wearing a black robe with a hood. Her robe ran all the way down to the floor which made her look as though she was floating across the floor. As they walked closer and closer towards each other, the young man became more and more cautiously afraid. But as they were about to pass each other in the darkness, she stopped and turned towards him. She told him she was from Pleiades, and that she was one of seven sisters. The strange woman went on to tell him, that as soon as he had passed away, there was going to be the start of WW III, which would destroy earth and all its life forms. But she provided him with a formula that could save some of mankind if not all. The formula she provided went like this;

the pituitary gland that would provide the special traits people have. It identified humans as to whom they are. It provided the energy to the human life form. If harnessed, it could provide special abilities to the human life forms.

When the human life form became ready to transition from its shell (the body), the pituitary gland engaged and joined the spectacular array of energy throughout the galaxy and continued its journey through time and space. $T^2 + S^2 = d^2$, where (d) equals dimensions. The pituitary gland when stimulated could open portholes in $T^2 + S^2$, and could open dimensions or portholes and allow you to see the other side of that dimension. Depending on the $T^2 + S^2 = d^2$, one opens, is what one will see. He was awakened when their plane hit some turbulence, but he continued to think about the dream he just had. He tells himself this dream is too crazy and would never come true. He had no idea what that woman was talking about. Besides, where is Pleiades? I never heard of this place, he told himself. He remembered saying to himself, by the time they send me to Vietnam, the was would be over, but look at me now. He wanted to go to war and learn what war was all about. Now those dreams were shattered. He went home as a proud marine for ten days of leave. When his leave was up, he reported to an infantry training regiment (ITR). All marines go through infantry training no matter what job they end up with. He ended up running K Company. There they learned how to toss grenades, fire bazookas and machine guns, and do night fighting.

One day, they had to compete against the other companies; each company was divided up to compete in different events. He chose to run through the mountains of Camp Pendleton. Since he could run like a deer, he figured that would be the best place for him. He learned something very important that day. He learned that what you see is not always what is real. You see, when they started their run, they took off like bats out of hell. This was what he liked, and we ran and ran through the hills in formation. Somewhere out in the boondocks, they came face-to-face with their competition. However just before they made contact, their platoon instructor had three guys fall back, and one acting as though he couldn't run anymore and the other two were holding him up, while he had the rest of the platoon marching. The other platoon thought they were done. But as soon as they turned away and were out of sight, they gathered up and started to run again. they don't know why, but the pace was much slower. They were all begging their platoon leader to go faster, but he kept on an even keel. When they got close to the finish line, he stopped them, lined them up in a good formation, and they marched their way across the finish line. They lost the competition by a couple of seconds. This trick was used later by the young man in the corps when he became a platoon leader. He used the same tactics, but this time, he made sure they won.

Chapter 14:
Be careful of what you wish for

After completing ITR, he reported to Camp Del Mar where he underwent admin training. he hated it; it was not something he wanted to do. In any case, he graduated in the top ten of his class. After graduation, he went on ten days of leave. He made the best of those ten days. He was with his girlfriend. Remember her, the one with the big green eyes? And of course all the girls at home liked him. It was the uniform he had to thank for all the attention. He had a great time at home. He must have because when he left for his duty station, which was at North Carolina, he didn't have a dollar to his name. He left on a commercial flight; it was a direct flight from Los Angeles to Dulles Airport. He could recall it was winter because when he arrived at Dulles, he was very cold. It was so cold that he doesn't remember how he got to North Carolina. He thinks he took a prop to Piedmont, North Carolina. It was the worst flight he has ever had up to this point. From there, he took a bus to Jacksonville, North Carolina. When he finally arrived on base, his unit was gone, so he reported to a recon unit next door. He spent around three nights with them; they asked him if he wanted to stay with them. He said, "No way." he thought to himself, these guys must do push-ups in their sleep.

His unit finally arrived. He reported in, and everyone got settled in. That very first night, some sergeant came to get him out of bed and told him he had guard duty. They drove him out to the boon docks to guard the motor pool. There was snow on the ground, and he was freezing. They dropped him off just before dark. As darkness came, he started seeing these little flying bugs that lit up. That was the first time in his life he ever seen a firefly. He couldn't help himself; he left his post and started capturing some. He placed them in a little plastic bottle he had in his first aid kit. He was just in awe to see all this light illuminating from the tree line. They looked like Christmas lights. While on guard duty, he encountered a giant snake crossing his path. He figured he would hit it hit with the butt of his rifle and kill it. the thing. He hit it as hard as he could, and his rifle bounced back as though he hit a tire. The snake stood up and stared at him while making this hissing sound. He just stood still, and the snake went down to the ground and proceeded on going to wherever it wanted too. It went straight into the motor pool.

As the night went by, it must have been around two in the morning, he reported that a snake went into the motor pool. He wanted the men to be careful when they climbed into their vehicles. Now being in North Carolina was like being in a different planet. It took him about three days to begin to understand what they were talking about. They had a completely southern drawl: words like up yonder or down yonder and words he never heard of before. But before he knew it, he was talking just like them. This was where he learned what prejudice was all about. The whites didn't like the blacks and vice versa. He was lucky he could hang out with either side. When he was with the whites, they would see four or five blacks together and say "Look at them negering over there." They would tell him that back home, they would have these runs where they would go shoot up the black neighborhood and how they would catch one and hang him on the nearest tree. He couldn't believe his ears. He told himself, these guys must be nuts. When he would hang with the blacks, they would confirm their story. So the blacks hated the white men just as bad. Here they were—having to live together and work together as though there was nothing hateful going on. Their barracks were next to the enlisted man's club and right next to the club was the women marines' barracks. They couldn't have asked for anything better. One night, he was standing in front of the club, but stood closer to the BAM (broad-ass marine) barracks. He was waiting for one to come out so he could invite her for a few beers. Oh, that's another thing: you could drink at eighteen over there. That was a big surprise to him but a good surprise. Well, there he was, smoking and waiting for a girl to show up. He didn't care what color she was.

When two black men showed up, acting cocky and bad, they demanded he give them his pack of cigarettes. He told them no. They stated that if he didn't, they were going to kick his ass. Well, he immediately pulled off his belt and whacked them across the face with the buckle side. They started hooping' and hollering and started to run. Well, he chased them with his belt to the back of the club. The young man slipped and fell but immediately got himself backup. By then, the entire club had emptied out and was surprise to see he beat them both up. It turned out they belonged to another unit. They were there just looking for trouble.

The very next day, he was the main topic of conversation— "that a little clerk kicked two guys' asses." An admin class was just starting up, so the first sergeant decided he should go and get retrained because the East Coast did things different than the West Coast. So off he went to Admin School again. He did great; it was basically the same thing—just some wording was different. He graduated third highest in his class. It made the skipper, first sergeant, and the admin chief very happy. He was promoted to private first class, and life went on. He still wanted to go to Vietnam, so he started putting in for transfer to WESTPAC. But they didn't want to cut him loose. He was good at his job, so before he knew it, he was promoted once again, this time to lance corporal. But he was still putting in his transfer request. It got so bad that the captain finally called him into his office. He asked the young man if anything was wrong and if he didn't like it there? The young man replied, "Sir, I joined the corps to go to Vietnam. Instead, I was made a clerk and sent here. I really want to go to Vietnam, sir. That's why I joined." The captain replied, "Son, a lot of people are working hard not to go there, and you're working hard to get there. I just don't understand it. But be careful of what you wish for."

My Bestfriend and I

Well, he made a good friend there; his name was Herman Raymond Lee Smith. For some reason, he will never forget his full name. They used to go on weekend liberty together. They would pay someone five dollars each, and the guy with the car would fill his car up with marines going his direction. They used to go to Washington, DC, a lot. One night, they met this hustler; he was a black man. He told them he knew of a place where they had lots of women and plenty of booze to drink. So they followed him to this hotel. As they were climbing the stairs, he told them they should put their money in a safe place because the girls would rip them off. So Herman Raymond Lee Smith did as the hustler requested. When we reached the top of the stairs, he asked the young man for his money. Again, something didn't feel right, and he refused to give him his money. The hustler talked to Herman Raymond Lee Smith again and told him to tell him to have him place his money in a safe place. Smith finally told the young man, "Look, I'll give you your money back if we lose it." So like a big dummy, he gave him forty dollars— that's all he had. The hustler put all the money in an envelope and walked over to Herman Raymond Lee Smith pulled out an envelope from his coat pocket and had him sign his name on it. The hustler told Smith, "Look, when you go in there, give them the envelope and tell them to put it in the safe for you." The young man stayed back, and as the hustler was passing him, he grabbed the hustler and stopped him and asked Smith to open the envelope to make sure their money was still there. Smith responded, "It's all here. I can feel it." So Smith and the young man were arguing then. He was asking Smith, "Just open the envelope," and Smith replied, "It's all here. I can feel it." His famous words were, "Don't worry; I'll pay you back if we lose it." Well, he let the guy go, and off he went down the flights of stairs. As he walked toward Smith, Smith opened the door. The place was empty. He finally opened the envelope. There was no money—just cut-up paper the size of dollar bills. They were now broke with no money to get back to base. The young man ended up making an emergency phone call home and asked his mother for forty dollars sent via Western Union.

He received the money the very next morning. So now we had enough money to pay the driver to take us back to Camp Lejeune. Each payday, he would ask Smith for his money, and he wouldn't give it to him. One payday weekend, they were out in town, having a few beers, and he asked him again for his money. Again, he refused to give it to him. Smith stated since he was part of the scam, he shouldn't owe him any money. He may have grown up poor, but where he grew up, a man's word meant everything. Without your word, you're nobody. He tells him, "What kind of person are you, not keeping your word?" So the young man tells him, "Let's take this outside." They walked into an alleyway and started fighting. He was pouncing on him. They were on the floor, and he was on top of Smith, just letting him have it. Just then, someone grabbed him by the shoulder and told him to stop. It was the police, the officer talked to them, and they related their issue to him. He asked if both of them had enough. They both said yes they did, and he let them go. They went back into the bar and had some more beers.

But the young man was satisfied because he felt he got his forty dollars' worth of kicking his butt.

Then the day finally came. He received orders to report to FMS, WESTPAC, for duty. Yay, he was finally going to Vietnam. So the day came. He packed his sea-bag, said his good-byes, and off he went back to Camp Pendleton, California, plus he got to see his girlfriend once again. This was sweet. Things just couldn't have gotten any better. He arrived home with ten days of leave again. Everyone was commenting on his speech. They said he was talking funny. Well, that was because he just came back from a redneck state with the kind of people that still fly the Confederate flag. Being home, it didn't take long to start getting his regular speech back. So there he was back at Camp Pendleton, California. They put him to work right away. They were short of admin clerks. So there he was in the middle of a processing center, processing marines in and out. Before he knew it, guys were going over to Vietnam, and they were just keeping him there to help them out. They didn't want to cut him loose either. He finally had to go see the lieutenant and talk to him about his situation. So they took him off the work detail and put him in with the guys going overseas. He finally got his orders. He was to depart on December 15, 1966, for overseas duty. The young man figured he joined August 30, 1965, and now he was finally going to Vietnam.

He arrived at Kadena Air Force Base, Okinawa, and transported to Camp Henson, Okinawa. There he was placed in a staging area where they were building up the force to go to Vietnam. He was there for about ten days. Suddenly, there was like battalion strength of men. They called everyone out on formation; there were a lot of men ready to go over to Vietnam. They called out just three names; one of those names was his. They were to report to Headquarter Ninth Marines there on Okinawa. He couldn't believe his ears. Everyone else went to Vietnam except them three. He was good at what he did, so they wanted to keep him there, and they did. He made his home there, and soon he was promoted to corporal. He liked it in Okinawa. He liked the way the people lived. It reminded him of home, living in the garage. Their houses were small but neat and very clean. And the girls, well what could he say, they were just beautiful. He even learned a little Japanese. But this still didn't sit well with him. Because he still wanted to go to Vietnam. So again he put in for transfer to a unit in Vietnam. They ended up giving him his request. So, he reported to the transition area at Camp Schwab, Okinawa. Someone put him in charge of everyone coming into the transition area.

Chapter 15:
Marines' Life

In the universe there are portholes all around everyone. People just need to learn how to become seers. So, he jump through one porthole and became a leader. He would call everyone out in formation and give them assignments that needed to be done each day. Once they had enough men he would appoint squad leaders. They were to make sure whatever duties their squad received would get done correctly. Things came to him in dreams, he didn't understand why. But he had this dream of being in front of this pound of water with these bushes in the back of the pond. There were two Korean marines standing on each side of the green plants; they were pointing to him that someone was in there. He knew he should have used his grenade. However, he figured if someone was in there, he would end up shot dead. So, since his rifle was pointing in that direction, he just unlocked pointed it in that direction and emptied seventeen rounds into the brush. The staff sergeant in charge came out madder than hell. He asked, "Who fired those shots?" the young man said, "I did." He came over and grabbed him by his shirt with both his hands and said, "You stupid asshole. You just gave away our position!" He continued on and told him to go up this knoll and keep his eyes open for enemy troops. The company came in a force march, not knowing what they stumbled into. He can recall the captain standing in front of the pond of water and asking the staff sergeant where the blood was coming from. "Oh, it's just a dead carcass," he answered. That really bothered the young man because he didn't know if he killed children, a family, or enemy troops. He guessed he would never know. But knew he hit someone or something.

The unknowing would go on for days. There were days where they didn't make contact with the enemy, but they had to move through 110 to 115 degrees of heat. There was no hot chow for them to eat, just C rations. No clean water to drink—they had to drink the water found out there. They used to write home and ask for Kool-Aid packages. That way, the water didn't taste so bad. Oh, by the way, you know that hero stuff he talked about as another reason for coming to Vietnam? Well, that went out the window the first day out in the bush. It was nothing but survival from that point on. He was asked if he wanted to go on R & R (rest and recuperation). You see, you got an extra R & R if you captured NVA officers, but he was so gun-ho that he turned it down. Big mistake, but he turned it down just the same.

One day, they were ready to settle down for the night; as he was going to set up the LZ for their night and morning chow drops. He had two VC prisoners to place on the helicopter, and that was it. However, as soon as the first helicopter landed; they were hit with an M79 grenade, which is a grenade launcher. It landed right in the middle of them all and close to the helicopter. All they were able to unload were two boxes of chow, and the helicopter took off. At the same time, the column of men up front got hit with about five NVA solders; they had an AK-47s, a machine gun, and a M79 grenade launcher. Bullets were flying all over the place. They could be heard as they wizzed by everyone. The young man moved his men into a cement structure; it was a temple for the dead. So bullets were flying in all directions and they were ricocheting all around. He looked over to the LZ, and saw one of his guys grabbing his throat, gasping for air. With bullets flying overhead, thinking the marine was hit in the throat, he ran up to the marine. When he got there, he found out that was not the case. The marine didn't want to leave the two cases of chow out there. The marine was trying to bring the boxes back. He slipped and the box hit his throat. He tells him,

"You stupid fool. You got me out here for this, I ought to shoot you myself." They got back to the cement building and decided they would fire back although they couldn't see anyone. So they laid out suppressive fire; one guy would unload his magazine out the window, and he would do the same while the other guy was reloading. This went on for a couple of magazines each, and the firing stopped.

In the meantime, the point man up front of the column took out the machine gun and the five NVA soldiers who had set up the ambush. He was wounded in the stomach by the M79 grenade launcher, from the NVA up in the trees; which he also ended up taking out. They also started receiving fire back at the LZ again. So they started reporting to the choppers that they had a man with a stomach wound, but they also have a lemon LZ. So all the choppers started to leave the area. Then all of a sudden, a chopper just popped out from the tree line. While under fire, they were able to put the wounded man into the chopper, and off the chopper went. Some other marines took care of the NVA's firing into the LZ. Then all the shooting stopped. The wounded marine received a Silver Star for his actions. So did the guys in the chopper because they weren't required to enter into the hot LZ.

One time, they were high up in the mountains, in the middle of a lot of trees. They got into a firefight, and they had one guy wounded. The choppers couldn't land to transport their wounded marine because of all the trees. So, they dropped a basket while they hovered over the trees. They strapped the wounded marine into the basket, and off the chopper went. Shortly thereafter, the unit received a call from the chopper, stating they lost the basket with the wounded marine. Their skipper ordered a search party to go find him. Upon the return of the search party, they reported the bad news. It looked as though the basket got stuck in some trees. They were unable to release the basket so they shot the already wounded marine and cut the rope to the basket to free themselves from the tree. Their first sergeant went crazy. He was asking if anybody got the number of the chopper because he was going to shoot them down himself. That's how it was in that crazy world. You couldn't trust anyone. From time to time, they would go to staging areas and be picked up and taken back aboard the ship. They would be aboard ship for about three to four days, and off they would go by helicopter inland again on search-and-destroy missions. The navy sure took good care of them while on board; They had good chow to eat, showers, and good sleeping quarters.

Every time they would kick off an operation, they would get up very early while it was still dark. The navy would feed them well, anything they wanted. It was like a last meal before being put to death in a prison. For many of them, it was their last meal. They would all line up, and the navy would help them load on the choppers. They were using the old choppers; they could only fit five fully combat-loaded marines at any one time. The navy would give them a piece of paper, and they would put their last name on it. The last guy, or the fifth guy, would hand over the slip of paper to a navy guy, in that way if the chopper went down there was a record of who was on the chopper.But the last guy with the slip of paper was the last to load, so he had to sit by the open door near the door gunner. They would sit on their helmets just in case they flew into a hot LZ, so they wouldn't get their butts shot off.

The young man recalls the first time he went on an operation, he noticed people jockeying around so they wouldn't be the man with the paper or the fifth man. And him, like a big dummy, just stood where he was, and he ended up with the paper. He was the fifth man, so he had to sit by the open door near the door gunner. The choppers would all be cruising at a certain altitude, and by the time they got over land, the sun would be coming out. As soon as they reached the LZ (landing zone), they would drop from the sky and hover above the ground two or three feet off the ground, and they would have to jump out of the chopper with about thirty to fifty pounds on their backs. He could remember when the chopper was descending down to the LZ; he was descending so fast that he thought they were shot down and that he was going to slide right out the door. He could tell you it was some type of an E-ride; it really got your heart pumping, and you see the last guy with the paper is also the first guy to jump out the chopper. So he put his helmet on, and as soon as the door gunner said "Go," he jumped off the chopper. He was fortunate that day because they landed on a sweet LZ. So the skipper gave the platoon leaders their assignments and directions to travel down into the rice paddies. One platoon was sent to the left flank, another to the right flank; one platoon took point, and the other platoon took the rear. The young man traveled with the skipper, first sergeant, and a handful of men; some were from the first shore party. He was responsible for about eight men. He never knew or learned what the men did, but they were with him. They would help him set up the LZ when one was needed. With the young man were a raddio operator and a corpsman.

One day, they came across an un-exploded bomb from one of their planes. So, the skipper decided to disarm the bomb; to prevent the enemy from using it as a booby trap. Well, the demolition guys made a mistake, and the bomb, a high-explosive devise, went off. All that was seen were pieces of meat fly across the air and landing in different directions. It was a high-explosive device. You know, after a while, you just didn't care if you died or not. You told yourself, if there's a bullet out there with your name on it, then let it be so. Once you came to accept death, you actually relaxed. You weren't bothered with the though of being killed anymore. By accepting death, you were no longer afraid of anything. What could happen? What is someone going to kill you?

Then we transferred over to the USS Iwo Jima, while the USS Okinawa went to the Philippines. It was still the same old thing, just a different ship. He remembered when he came down with what was called rotten crotch. His crotch was bleeding and raw. There had been a firefight that night, and two men had been wounded. They decided to evacuate him out with the wounded; he was tagged. He remembers landing aboard ship. The Navy personnel came out to them and immediately started taking their weapons, and clothes off. By the time they got to the hanger deck, the wounded men were completely naked. The hanger deck started descending, making a loud noise or whistle sound. When it stopped, the wounded were put on stretchers, and doctors were ready to attend to them. The young man was told to go get cleaned up and put this cream on his crotch and his entire body. This took place in the end of November because on December 10, 1967, the young man's company was hit with enemy artillery. They lost close to a dozen men that day. He felt so bad and was so sad for the guys that got killed. He felt he should have been out there with them. Instead he was aboard ship, living the good life. Maybe things happen for a reason. He guessed he was lucky being aboard ship that day. But this was when he had to go to work as the Admin Man. He had to type the letters of notification to the deceased soldiers' mothers or spouses. The letters had to be perfect—no errors, no erasing or the use of White-Out. The requirement of something like the original and seven carbon copies was needed. There are a lot of letters out there with the initials RAR in the reply section of each letter.

Back to September 1967, before the young man was medevac from the field; there was this young black man, Otis Smith, who was new to the unit. He befriended the young man right away and asked him to take care of his special dollar. Otis was a big, strong guy. However, somehow Otis knew he wasn't going to make it. The units' corpsman also befriended the young man and taught him everything about giving someone first aid. He showed the young man where everything was located in his bag. The corpsman also showed the young man how and where to stick him with the morphine in case he got hit. Every marine is taught to provide first aid. However, the corpsman was making sure the young man knew as much as he did. So, the corpsman and the young man talked about first aid as they walked across the rice paddies of Vietnam.

Fortunately for the corpsman, he never got hit, so the young man never had to use what was taught to him. Now Smith was another story; he was new to the unit and was only with them for two weeks before he was killed by a sniper's bullet.

The young man can recall it was a very dark night, you couldn't even see your hands in front of your face. It was raining hard, and they were spread out thin because they were short of men—one man per foxhole. Well, it was decided to put out a listening post right in front of the young mans' foxhole about twenty-five to fifty feet out. About five of men went out to the listening post. Sometime during the night, He hears two shots. He also sees the muzzle flash; He gets into position to fire at where he saw the muzzle flash, but the lieutenant came by and asked The young man to keep his head down. Suddenly, a radioman came down, and the lieutenant was talking to someone out at the listening post. The lieutenant was informed by the listening post that when they tried to return fire, all their weapons jammed. The lieutenant called them all back in. A corpsman was already out there. The men all came back, including Pvt. Otis Smith's body. They placed his body right behind the young mans' his foxhole, and there the body remained all night. with Private Smith's body. The young man couldn't believe Otis was dead. He had just seen Otis walk by him to the LP. So he gets out of his foxhole to move Pvt. Smith's leg to see any sign of life. He even got out a second time and went directly to his head, placed his hand behind Smith's head and his ear to Smith's lips; all he felt was this sticky and gooey stuff in the back of his head. The man was dead. He was with him all night long. At least he died instantly, he didn't suffer. The young man asked other marines to bring him two large silkworm disks to place under and on top of Smith to protect his body from the elements.

In the morning, the corpsman came by to wrap and tag him as a KIA (killed in action) for the report. The corpsman used Smith's bootlaces to tie the poncho around Smith. He wrapped one bootlace around his ankles and the other bootlace around his neck. The corpsman noticed something odd about his death though. He noticed powder burns on Otis' face, meaning that he was shot at a very close range. The young man thought the corpsman wanted him to include this in his report. But the young man figured that was the corpman's job, not his. "You start making accusations like that out in the bush, and you might not make it out yourself, he thought to himself." The South was mighty powerful out there. One didn't know who was going to kill you, so you really had to watch your back. It took four guys to carry him out to the chopper that morning. That morning, he had to call in the morning report in code: "Number four, "1", Michael, India, Tangle, Hotel. Did you copy that!" He had about ten items to report on each morning; number four was the number of KIA's, and he provided the last four letters of their last name.

Chapter 16:
Life in Battleships

One afternoon, they contacted the NVA; the enemy had AK-47s and a machine gun. The enemy was returning heavy fire and was well dug in. The young mans' company had to call in an air strike; an F-4 showed up, carrying napalm bombs. The F-4 flew over to look at their positions. They were all lying on the ground because bullets were lying everywhere. The company had their red air panels out so the pilot could see their side of the line. The pilot made a pass and looked over the situation, and then he started his run. The young man looked back, and he could clearly see the F-4 was coming right at them. When the F-4 released two napalm bombs; the bombs looked like they were coming right at them. He could see the two huge bombs tumbling through the air. He tells himself, hope The pilot didn't make a mistake in his judgment. The bombs tumbled over their heads very low but kept going and landed on the enemy. He remembers when the bombs went off; the whole ground shook, and it seemed as though all the air was sucked out of the area for a few seconds or so. This action broke the enemies backs. The soldiers were able to charge the trench and eliminate the threat. The NVA had Chinese and Russian advisors. At times, they would end up being caught and mainly killed. This had to report to higher command. The young man figured they had some type of protocol to advise the Chinese or Russian government that their son was killed in action. Yes, they used to run into a lot of highly trained Chinese troops. The Chinese were bigger and taller than the NVA, but they also displayed a star on their belt buckles. The young man guessed the Chinese would fight to the death because he never had a Chinese prisoner of war to look after.

One day, they were done with an operation. As they were headed to a staging area to be picked up by trucks, they had to cross this stream of water that was about waist high. After they crossed the stream, they were near the staging area. One of the marines tripped a booby trap. It was a grenade in a can. Well, he set it off. The young man saw the explosion and about five men falling backward. It looked as if they were falling in slow motion. He was reminded of a flower opening up. All five were wounded by the blast. However, one marine got a piece of shrapnel right in the heart. A medevac chopper was called in. The corpsmen tried very hard to keep him alive, but by the time the chopper arrived, he was dead. The first chopper to arrive took all the wounded, and the second chopper took the dead marine. They finally made it to their extraction point. It was weird; there were lots of civilian Vietnamese to greet them, and they had cold beer, sodas, cigarettes, and chocolate candy bars to sell them for a buck a piece. It was a bittersweet moment; here they were coming out of the bush, and the civilians already knew where they were coming out from. The marines didn't use dollar bills or any type of American currency; they used what was called MPC money. The young man no longer remembered what that actually stood for. It was something like "military pay coupons." or something like that. But as soon as they got on the trucks, they started to pull their pants down because they all had these big, giant tiger leeches all over their legs, their manhood parts and some on their torso. It all depended on how deep you got into that stream of water. Some of the guys would use cigarettes to get rid of the leeches. Others, like himself, used mosquito chow to squirt on the leeches, and they would unscrew themselves right off your body and blood would run out. He also remembers having lice; they were crawling all over his crotch.

They finally reached their destination, and from there, they were picked up by helicopters and taken back aboard the USS Iwo Jima. He immediately reported to the corpsmen and told them of his problem. They gave him some cream and told him to shower and use the cream from head to toe for about three to four days. It worked. He got rid of all the lice from his body. It wasn't long before we were transferred to another ship, one that had landing craft aboard it. In the morning, they saddled up and boarded these landing craft. They were some type of Amtrak because they were fully covered in steel. They climbed in through an opening at the top and went inside. There were some eight to ten guys, fully combat-loaded marines, inside each Amtrak. He could remember seeing those giant doors opening up on the ship and the ocean water coming in. There they were crossing over the ocean toward the beachhead. They started hearing enemy rounds hitting their Amtrak.

One of their men started firing back from the top of the craft. While they pushed farther inland, it stopped, dropped the door open, and they all ran out, taking positions. It was a short skirmish, and the company pressed on inward. As often happened, they came across a village. Their orders were to check it out and to determine if they were helping the NVA or VC. The evidence showed that they were helping the enemy; so, they received orders to burn it down. So they moved on, leaving the village ablaze. The young man could see the women and some children running to save their homes. So they called in an artillery strike and just leveled the place to nothing. That's what they used to do when they would find villages helping the other side. That was why they were on search-and-destroy missions. Any prisoners captured would be given to him. He would blindfold them and tie their hands behind their backs. He would guide them along to their destination, the LZ, so they could replenish their supplies and ammo and drop off any new men coming into the company. He remembers one POW was so scared; his heart was pounding right out of his chest. They were traveling through rice paddies full of water, and one of the marines came to get this prisoner to put him out in front of him. This way, if they ran into a booby trap, the prisoner would set it off, thus saving the marines life. He never got that prisoner back. The young man never got the prisoner back, nor did he know what became of the prisoner. Perhaps the prisoner died of a heart attack or something. The young man used to try and take care of the POW's. He would give them food and water while they still had their hands tied up behind their backs.

As they were settling down one night, there was this terrible smell. It made you gag. It was so bad, it made you sick. All of a sudden, He hears his name again. "Ramirez, get two guys, find the body, ID it, and rebury it." He gets two of his men to go with him. Of course, you had to use your noise to find it. The stronger the smell, the closer you got to finding the body. They found the body; they got to about three feet from it. They couldn't have gotten any closer. The stench was so bad; they were all ready to puke. So he just called it. "It's a dead gook, so let's bury it real fast and get the hell out of here." So there they were throwing more dirt on the shallow grave site until they couldn't smell it anymore.

One day they departed the ship and reported to a base called Hue; they were near the Da Nang area. They built them some billets with canvass tops to be lined with plastic siding so you could see out. The young man figured he was responsible for all the company reports that needed to go out. This was where he met Sergeant Zalenski; he was one of that redneck, confederate flag-waving marines. His office was right next toSergeant Zalenski's. All the other companies had to report to them. as well. The young man's skipper and company had joined with the other companies out in the field, continuing with the same mission: search and destroy. The company commander wanted the young man to stay there to help type letters to the mothers and wives of the twenty marines that were killed the day before. So, there he was in a place he didn't want to be. Zalenski, would shout from his office, "Poncho," meaning the young man, and shout, "Get your ass in here." When the young man reported to Sergeant Zalenski he always had to report at attention. Zalenski liked having that power and authority. He had a lieutenant that was just like him. You would think they were made from the same mold. The Sergeant would say "Where is my copy of your unit diary and my copy of your morning report?" The young man would tell him, "we are working on them right this minute, and you will have the reports as soon as they are done." That wasn't good enough for the sergeant. He told the young man to get his ass back and bring him those reports. The young man told him, "As soon as we get them done, you'll have them." "That's not what I want to hear. You get your ass back over there and bring me those reports." So he went back, finished the reports, and took the sergeant his copies. So, this was how life was in the rear area. The young man hated it! He didn't want to deal with people like him. He would rather face death.

Well, the day finally came when the young man had enough. Zalenski yelled out, "Poncho," and he replied "What do you want, Ski?" The sergeant yelled back, "What did you call me?" The replied again, "Ski." The segeant said, "Get your ass over here." They went in front of the man, their lieutenant. The sergeant said he wanted to write the young man up for being insubordinate. He demanded his strips. The lieutenant asked what went on, so the sergeant replied that the young man had called him "Ski" instead of addressing him properly by calling him Sergeant Zalenski. The lieutenant asked the young man, "What do you have to say?" The young man's reply was, "Sir, I have asked him to stop calling me "Poncho, my name is Corporal Ramirez." The lieutenant asked the young man to leave, so he left. The next time the sergeant called the young man, he yelled out "Corporal Ramirez." All the men in the office gave him a thumbs-up because they knew he stood up to him and got the lieutenant to square him away. That's when he decided to go back out in the field with the company. He just couldn't stand being in the rear area; to him it was not productive.

The young man remembers getting a letter from his mother. She told him, "Guess what? You have a baby brother." That really made him feel good. She went on and said his name is Jimmy; this was in 1967. So he stayed out there in the field until he was medevac out of the field with that bad rash between his legs. That was in late November 1967.

The ship had come back from replenishing itself in the Philippines. The young man remembered on December 10, 1967,his company was hit by enemy artillery, and lost about a dozen men. They even lost one of their corpsmen; he took a direct hit while being in his foxhole. He felt bad for the men. He felt as though he should have been with them. But who knows in the big scheme of things; just maybe it was meant for the young man not to be out there. That was why it was so important for everyone to do their job in order to stay alive out there. He remembered being out there and hearing about the Delta and Charlie companies being overrun. The enemy got passed the listening post and broke through a weak side of the line. They just went right in there, killing the skipper, the exec, the first sergeant, and all the men in the middle. The enemy was a hard-core fighting machine; you just couldn't do anything wrong out there because the enemy would exploit it and take advantage of any given situation. He could recall being out in the field after Private Bush was killed, and he spent the night with the body.

As they set up for the night with two men per foxhole. The young man remembers seeing this flickering flash coming from the foxhole next to his. So he thought he would go and check it out. It was a brand-new second lieutenant with his bars on. The lieutenant's bars were giving away their position. The young man explained this to the second lieutenant and he immediately took off the bars and said thanks. He goes back to his foxhole. Sometime during the night, he started hearing someone whispering his name. It was a real faint whisper. It was their guys that were assigned to stay out in front of the company as a listening post. They begged him to let them come in; they were saying they were going to get killed out there just like Smith. He lets them in, and he and his partner hid under the bushes and slept all night while they took their watch. Now you see how things can go wrong. He weakened that side of the company; they no longer have the ears out there to alert them of incoming enemy troops. He did that once and never again. Those two men snuck back out in the early morning and went to their assigned position and called in, saying they were coming back in which they did. They were lucky that night. His Vietnam tour was up, and he received orders to report to Camp Pendleton's rifle range. He was an expert shot, so the rifle range would have been perfect for him. Sergeant Zalenki was going crazy; he wanted the young man's orders and was trying very hard to take them from him. Well, he ended up handing over that assignment to him on a silver platter because he put in for an extension to his Vietnam tour just to be with his company for another tour.

In the meantime, they shafted him. His extension was approved. So now they had control over him once again. So they transferred him to Alfa Battery 1/12. He was told this was temporary until he started his thirty days of leave anywhere in the free world. He chose to come home for thirty days. In the meantime, he became the admin man for Alfa Battery 1/12. He was by himself, and he learned that the prior clerks didn't keep up the records. So there he was, working from morning till late night trying his best to catch-up all the records by checking every aspect of admin duties. The battery was out in the field, so there he was all by himself. One time a lieutenant came by, as he was going on R & R to meet his wife in Hawaii. He left instructions with the young man regarding taking care of his.45-caliber pistol and holster. He wanted his holster and weapon returned to him upon his return from R & R. The young man told him, "Sir, I am going on thirty days of leave, and I won't be coming back. I am here only temporarily until someone comes in to replace me." The lieutenant said, "I don't care. I want you to make sure my pistol is here when I get back." Well, when it was the young man's turn to leave, they sent over a non-admin type of person just to cover the office. He gave his replacement specific instructions about the pistol and holster. Then he also gave him paperwork from some of the men to increase their allotments to their wives. He tells him, "Make sure these go up to battalion for processing." After explaining the instructions to his replacement, the young man takes off on his way home.

Chapter 17:
Seventy-seven-day siege of Khe Sanh

▲ A U.S. Air Force C-123K transport was hit by enemy fire over Khe Sanh on March 6, 1968, killing 50 Americans.

When he arrived at Okinawa, he was asked if he wanted to make it home for Christmas. He tells them, "No, I'm not in a big hurry." Besides, he had a girlfriend on Okinawa, and he wanted to spend some time with her. So he spent around three days there with her. Then his flight came up, so he was on his way back to the world. When you were going back home; it was referred to as going back to the world. It was great to be home. He jumps onto his younger brother's bed and takes a nap. There he was sleeping on a bed with clean sheets and a pillow and eating good homemade cooking. When he awakes, he goes to the refrigerator, pulls out a cold beer, wipes his forehead, and says, "What a nightmare" One day, one of the girls next door wanted to go out with him. So, he bought two tickets to see this live concert. When he went and knocked at her door, her mom said she couldn't go. So, he gave her mother the two tickets and told her to give them to her daughter as souvenirs. Later he learned she was already pregnant. He was so glad the girl's mother decided not to let her go, and if one thing led to another, he could have ended up being the fall guy. She was a young good-looking girl, but he wasn't ready for fatherhood especially with it not being his kid. His stepfather was a cool guy. He got them all out of that garage in East Los Angeles. He would loan him his car so, he could go out and have fun. He ended up spending most of his time visiting his girlfriend. Well, before he knew it, his thirty days were up and it was time for him to return to Vietnam. He remembers his mother telling him to stay home, not to go back—that he already served his time.

The young man was very tempted to stay but he knew he would end up in lots of trouble. So he had his stepfather and his mom take him to the airport. He had to report to A Battery 1/12, so off he went and landed in Hawaii as his first stop. The plane refueled and took off to Okinawa. From Okinawa, he took a C-130 transport plane to Vietnam. Once in Vietnam, he learned what hill "A" Battery 1/12 was on, and he took a chopper to that hill. When he arrived, he was called into the lieutenant's office immediately, and he had to stand before him at attention. The first thing out of the lieutenant's mouth was, "Where is my .45 and holster?" The young man explaiined he left it with the guy who took his place. He said, "Well, I never got it back, and you were responsible for it." Then he got on the young man for the paperwork from the men. The young man told him that he told his replacement to send those allotment papers to command. However, the guy who took his place was not an admin guy. So, apparently, he just sat there doing nothing. Record books were not being completed during the time the young man was on leave. The lieutenant was really upset with him. He told the young man that he was a corporal in the Marine Corps, and that he should have been able to take care of all those things. The young man told the lieutenant "But I wasn't even here, sir." The lieutenant responded to the young man, "Get out of here before I take your stripes." But there the lieutenant was with four admin clerks, an admin chief, and himself. These guys really had it made here, sir." He told him, "Get out of here before I take your stripes.

"Khe Sanh Military Base"

The lieutenant was so mad at the young man that the very next day, the young man was transferred from the unit. He was sent to Khe Sanh. So there he was with his sea-bag on his way to Khe Sanh, Vietnam, the hellhole of the world.He had to report to A Battery 1/13 artillery unit.When he arrived, he couldn't see a soul, not a single person. He started walking up the road, not knowing where to go, when he noticed this helmet bobbing up and down this hole. This reminded him of a gopher, sticking his head in and out of a hole. As he walked down the road, he noticed all the housing was blown up and there were downed helicopters and downed aircraft everywhere. There were flipped-over trucks. For the first time, it looked as though we were losing the war.

Here are some pictures of the base below:

Khe Sanh

He found out later everyone was taking shelter in bunkers. Then this jeep came by, like a bat out of hell, and quickly stopped. The driver asked the young man where he was headed. The young man replied, "To A Battery 1/13." The driver said, "Hop on in. I'll drive you there." So he was dropped off, and he reported in. They were happy to see him; they made him feel welcome. The first sergeant took him around each gun and introduced him to the men. Little did he know that some of the men he just met that day would be dead by nightfall. The company started taking artillery, mortars, and rockets from the enemy that night. Gun one took a direct hit, which killed all three men inside the bunker; the rocket tore them to pieces, and a fire started inside the bunker. About three or four men ran inside to put out the fire. There was nothing anyone could do for the men inside.

The next morning, he was typing letters to the mothers and wives of the deceased. It was something like the original and seven copies with no errors, no white out and no eraser. If you mistype something at the very end, you had to start all over again. He had taken a typing class in high school. He was at about seventy words per minute, so he could type. That's why all the letters that needed to be typed ended up with him. He immediately knew that Khe Sanh was not a place he wanted to be. There was no place to go and no place to get away from all the shelling. They were surrounded at any one time by twenty to forty thousand enemy troops; they were well armed and well trained. Day and night, rockets would come in screaming at them. They could hear the rockets come out of the tubes and had a good idea as to where they were going to land, just by the loud noise they made. They were coming in screaming death. That lieutenant that sent him there actually sentenced him to death. The NVA had Russian and Chinese tanks that would fire there 90-mm rounds at them. They also had 122-mm rockets they would fire at them day in and day out. Nonstop they would throw these weapons at them. They were trying to take them out because their artillery could go out and reach their troops. We used to have a six-seater outhouse and a large hole to dump their trash. The rats there were huge and just multiplied; big, giant rats were everywhere. If a guy didn't get wounded, he got bit by a rat and would have to get his treatment shots right in the stomach. Life was hard there at Khe Sanh with no hot meals, and hardly any water, so no showers. Everybody was pretty funky. No one ever took their boots off because they were afraid of the rats biting their feet.

"Khe Sanh Bed"

One of the corpsman befriended him and started showing him first aid. The young man told the corpman he had already been trained by another corpsman, so he didn't have to worry because the young man already knew what to do. This corpsman wouldn't go to the outhouse without him. One day, he needed to go to the bathroom, and the young man wasn't ready to go. He was after him for hours, asking "Do you have to go now?" He tells him he wasn't ready to go; he was telling him he was going to shit in his pants if he didn't go soon, so he finally was ready to go, so he went to him and said, "Let's go." He wasted no time and led the way to the outhouse; there were four men in there. With them two, it made a full house, and there we were taking a dump when we all heard three rockets leave their firing tubes. We all waited because we had about seven seconds to take cover. As they heard the rockets screaming right at them, the other four men just ran out of the outhouse with their pants down, and they didn't even wipe their asse. They ran right into a trench that was made just for This type of situation. But somehow, the corpsmen and the young man decided to stay and save their dignity; the three rockets came and just slammed themselves right in front of the outhouse. It was tilted to one side, and perforated with holes. The doc (corpsman) and the young man just sat there, holding onto our helmets. Those two were very lucky that day.

On another day, the enemy was throwing everything at them: rockets, artillery, and mortar fire. Five guys got hit by a mortar; they were building a new bunker. Needless to say, they brought them into our bunker where our office was located. Just before gun one was hit, the young man and the other admin guy were walking to their bunker to get their ammo belts in case they got overrun that day. As they were walking to their bunker, they had to pass by gun one. When the direct hit on gun one happened, they were outside next to gun one when it exploded. The young man was thrown up into the air; it all felt like slow motion. As he was going up into the air, he thought this was the way people died and went to heaven. But then he felt himself coming back down. He remembers seeing the ground so vividly. He could see every grain of dirt; everything was magnified. He landed on the ground; His ears were ringing, and he had blood coming out of his left ear. He couldn't hear people when they would talk to him. He ran into the large bunker, holding his left ear. He could still remember the corpsman's look when he saw him. The doc's eyes got as big as fifty-cent pieces, and he was asking, "What's wrong?" The young mang told the corpsman he was bleeding from his left ear and he couldn't hear people talk very well. The corpsman said, "You'll be all right," they went on treating the severly wounded. When gun one took a direct hit. The 105-mm howitzer was just mangled to pieces, and there were other wounded men that needed help. Doc Mcgail was shouting for Doc Lucero, and Doc Lucero was nowhere to be found. Doc Mcgail asked the young man to go with him to gun one and help him with the wounded. The young man agreed to help him. All this time, rockets, artillery, and mortars were raining from the sky. Just as they were ready to run to gun one, the corpsman changed his mind about having the young man him help him out there. He said, "Stay back. I will call you if I need you."

Khe Sanh Vietnam 1968

This picture shows three 122mm rockets hitting Khe Sanh with the 4th rocket getting ready to rip anything in its path.

Most of the men were hit in the chest and had these big sucking air wounds. They put the men in this ambulance type of jeep that could carry a lot of wounded men on stretchers. The hospital was right next door, so they didn't have to go far. It was still raining with enemy fire, and the men were all stating, "Let's go, let's go." They didn't want to get hit again. One of the men had lost the back of his head; he was fortunate he was wearing his helmet. Well, after things settled down, the young man learned what had happen to Doc Lucero and why he was missing for a while? It turned out that one of the staff sergeants got wounded and ran into his bunker and took Doc Lucero with him, pointed his .45 pistol at him, and demanded he be taken care of or he would shoot him. The staff sergeant didn't want him to leave his side until the ambulance jeep pulled up and picked him up. Then Doc Lucero was able to move around and help others.

Things got pretty bad for everyone up there at Khe Sanh. One day they shot down a C-123 with forty-two marines in it. They crashed along the side of the hill. They all died in the crash. Three of those men were from their battery. Two were returning from the main hospital in Da Nang, Vietnam, from previous wounds, and one was returning from R & R. Everyone perished on that plane. Forty-two marines plus the flight crew all died in that crash. He just knew the enemy was celebrating that day; they scored a good hit. Besides, the enemy owned the hills all around them. To a marine, this was perfect; the marines had them right where they wanted them. Well, this went on day in and day out for seventy-seven days. The president wanted the hill held so, they did. They called it the seventy-seven-day siege of Khe Sanh. Khe Sanh was located in the most northern part of South Vietnam; they were there to stop the NVA and VC from getting to the south. But the enemy still managed to get down south by way of Laos trails. The marines hit them with bombers and F-4 fighter planes, and they still kept getting through.

Chapter 18:
Dear John Letter

One morning, we woke up to see this big, long trench right in front of their encampment. The young man believed it was called the gray sector. The marines had decided to assault the trench first before the enemy had a chance to assault them. One of the first lieutenants and a radioman were joining the mission. The young man thought this was a bad idea because the lieutenant had just gotten a Dear John letter three days prior to the mission. Basically, a Dear John letter is a letter telling you "I am dumping you for someone else" or at least "I am dumping you." He was fit to be tied and not in a good mental state to go on this operation. The first platoon from Bravo Company didn't come back alive—not one of them. It is this writer's contention that the first platoon from Bravo Company 1/26 didn't come back because the first lieutenant called in the artillery strikes on his own self, thus killing his own men because of that Dear John letter he received. The firefight was so intense that they didn't go out to retrieve the bodies until thirty days after the operation. It got so hot (from enemy attacks and enemy fire that was so intense) that they halted all landings and started to resupply them by parachute. There were five thousand marines and one thousand South Vietnamese marine troops on that hill against twenty to forty thousand enemy troops. Marines developed that one-thousand-yard stare near the end of the siege. They used to assign the young man on roving patrol most of the nights. He guessed because he came from a grunt outfit. He used to walk their perimeter and make sure the guys on post were awake.

One night, a trip flare went off in their perimeter, so the young man was called and told to get a man from gun one to go check it out. Well, when he arrived at gun one's bunker, he asked for one man to come out with him to go check the wire. Well, this guy was taking his sweet little old time putting on his boots and the rest of his gear. The young man told him if he wasn't out in ten seconds, he was going to write him up for not following orders. Well, he finally came out of his bunker, and he showed him where they had to go. Just as they started moving out, they heard a rocket leave its tube, and they were listening for it. It sounded as though it was coming right at them, so they both dove to the ground, melding to the shape of the terrain. The rocket hit the exact spot that they were going to check out. If that guy from gun one hadn't delayed; they both would have been in a world of shit. Because they would have been standing right there. His delay saved their lives, and the young man will forever be grateful.

One day, the enemy attacked and had a direct hit on the ammo dump. The entire hill shook; the plume of exploding ammo was traveling up into the sky. It looked like a nuclear explosion; the bombs were flying and exploding through the air. A big mushroom cloud appeared over the hill; tear gas was all over them. They had to put on their gas masks. Now everyone was short of ammo, food, and water; the water was rationed to one canteen per man. They were very funky, including their uniforms. It started raining, and the young man and another marine decided to use the rain to take a shower. They stripped and soaped themselves down pretty good; when they started to take the soap off it stopped raining. Here they were naked on top of the hill, begging people for their water. Finally, a couple of guys came through and gave them some of their water. It felt great to be clean; but they still had some soap on them, so they had to live with it. It was their choice, and they still thought it was a good one. The only bad part was they still had to put on their funky uniforms back on. But at least they felt clean. People started acting a little crazy and doing strange things.

One time, we were under a heavy rocket attack, and this guy climbed up on top of the command bunker and started dancing with no care in the world. Some of the guys were yelling at him to get down from there, and he just continued dancing all through the attack. The corpsmen shipped him out to NSA Da Nang Hospital for full evaluation. He never returned. The young man and another marine in the bunker started to play with a live grenade; they would pull the pin out and held the spoon tight and pass it back and forth to each other. One day, while no one was around, the young man took the grenade apart. He blew the blasting cap and put the grenade back together again. He told his friend what he had done, and they started to play that game again while everyone was in the bunker. The other marines in the bunker were all telling them to stop playing around when their hands collided. The grenade fell to the floor, and the spoon popped off the grenade. All the men panicked and ran out of the bunker. It was funny to watch because they looked as though they all went out of the bunker at the same time. It was a very small opening too. The young man and his friend just laughed and laughed. They ended up throwing the grenade away because they didn't want to forget and get it mixed up with a live one. Other guys were making booze. They asked them for their cans of fruit and sugar. He gave them all he had; in return, he would get a drink. They had made a good working still, and whatever came out of it was very strong. It appears the drink bypassed the stomach and went straight to your head.

One night, he thinks it was New Year's or some holiday. The men were firing off illumination flares; they were shooting them straight up in the sky. There was hardly any wind, so the flares were landing right back on their compound. The guys would all be chasing them and trying to catch the little parachutes. They were knocking each other down. It was like playing rugby. Before you knew it, we started receiving incoming rockets on our positions. Everyone took shelter. He stayed outside, helping gun one with their fire mission. He would give gun 1, the charge they requested, and at times, he would load the 105-mm artillery shell and slam it into the breach block, and we would end up having us an artillery duel. He would help the men out in gun one especially during night missions. He became a pretty good canon cock-er. 0811 was the job number. Then the air force had this Operation Niagara where they would fly over at a very high altitude and drop their bombs. They couldn't hear the planes, but they could hear the bombs raining down on the enemy; the enemy was close to them, so they could see and hear the bombs go off. The sky would light up on fire, and the ground would rumble as they dropped their load. Then the F-4 would come in during the day and would fire and bomb enemy positions while we watched all of this take place. He remembers seeing an enemy gunner shooting at the F-4; with an anti-aircraft gun, they could see the tracer rounds coming out of the gunner's entrenchment. The F-4 turned around for another pass at him. They could see the F-4 diving at the gunner's position, firing rounds at him, and then releasing a napalm bomb at his location, and they could see the cloud of smoke come up from the gunner's position. They all cheered and said he got him. Just then as the F-4 started climbing from his dive, the enemy gunner started shooting at him again; they could see his tracer rounds trying to hit that F-4, so we cheered again— this time for the gunner because the enemy showed he had some balls. But the F-4 wasn't giving up; he circled around once again and did the same thing: fired his rounds and dropped another napalm bomb at him. This time, there was no return fire, so the marines cheered on the F-4 and felt bad for the guy with balls. That air force operation broke the NVA's back, and they retreated into Laos; the siege was lifted, and the First Cavalry came in, mopping up the place. They painted their logo on the airstrip, and helicopters began to land all over our airstrip. Prior to their arrival, the CB's built them a big command post, a supply depot, and a place to serve hot chow right next to their unit. So, the army came in and started making hot chow for their men.

Well, the enemy wasn't quite done with us just yet. They started firing rockets, artillery, and mortars at the airstrip. Well, the First Cavalry unit ran to their helicopters and took off, leaving everything behind and never came back. The young man and a few other guys ran into their chow hall and took these huge pots of chow from their chow bunker. They did this while they were still under fire and rockets were landing all over the place. But to the marines, the hot chow was worth it. They took all their food to their unit, and everyone was eating hot chow for the first time since the siege took place. Some other guys in our unit drove over to their supply depot, broke in, and took as many uniforms as possible. They did all this under direct fire. Yes, the siege was lifted, but they would still get rockets, artillery, 90-mm tank shells, and mortars; they just got them less often. They were fighting against Chinese and Russian advisors. This was where the NVA was getting all their supplies. The young man told the first sergeant he needed to go on R & R. The first sergeant asked him, "Haven't you gone yet?" he told him, "No, I haven't been on R & R." The first sergeant asked him, "Where do you want to go?" The young man tells him, "Okinawa." The first sergeant replied, "Okinawa? Nobody goes to Okinawa for R & R. You must have gone somewhere on your first tour of duty." He continued with, "Don't you want to ride with us out of here on the convoy?" The young man replied, "No I don't. I want to leave know." The young man asked for about four days of travel to get to Da Nang. He told the first sergeant, "That's how long it's going to take your guys to get to your destination." Altogether, he got about ten days of R & R, which was about the norm for any R & R. The sergeant gave him his orders, and off he went and reported to the staging area. There were a lot of men all in trenches, waiting for their names to be called out. He was dressed like a grunt loaded for bear. He was also as dirty and as funky as one can be. When he heard them call out someone's name twice and no one answered the call, he jumped at the opportunity and said "Here" and boarded the chopper. It took him to some hill where he started his search for his best friend Tony Zavala, but he was nowhere to be found. So he took off to another hill, then another hill. He was about to give up when he noticed a chopper going to Camp Carol, a hill he hasn't been to. He walked into their cafeteria, hoping to see him since it was lunch time. He didn't see him, so he was giving up and on his way when all of a sudden; he hears his name being called. "Hey, Raymond" He went over to him and gave him a hug. Tony told him your all dirty and smelly. He said, "Come with me. I will fix you up." The young man got a hot shower with hair shampoo and real soap to wash his body. He also gave him a brand new uniform, and he somehow managed to get a bottle of booze. They drank most of the day, and then Tony gave him his cot to sleep on.

Chapter 19:
Courage in the Battle

The very next morning, they both went and had breakfast together. It was a fun day being with his old friend, his next-door neighbor at home at least when he was living in the garage. Well, it was time for him to leave. He tells him thanks for his hospitality, and off he went to Da Nang to catch a flight to Okinawa. When he was in Da Nang, he heard that A Company 1/9, Third Marine Division, his first outfit in Nam, was somewhere nearby. He went looking for them and found them just before it got dark. When he got there, they were very busy putting together all the personal effects of twenty-five marines they just lost in a firefight. That night, they also offered him a shower and provided him with some beers and gave him a cot to sleep in for the night. Well, it got dark, and everyone went to sleep. Sometime during the night, the young man heard incoming mortars. They sounded off in the distance, but a siren started going off, and everyone moved into a bunker right next door to their sleeping quarters. Well, he crawled in there too, but he couldn't stand it. He left the bunker and went back to sleep in the cot they had provided him. The guys stayed in the bunker almost all night except for one marine who came out right after him. The marine asked him, "Ray, are you sure you're going to be all right?" the young man replied, "Yes, I'm sure." I'll let you know if they start getting too close."

The next day, they went to get some breakfast, and off he went and reported to Da Nang for a flight to Okinawa. One night while waiting for his flight to Okinawa, there was a sniper shooting into their barracks. A sound came over the loudspeaker that they wanted grunts to report to the office. Everybody started to hide or scramble away somewhere to avoid going to the office. He just threw a blanket over his head and pretended he didn't hear anything. Well, they found him, grabbed him and took him into the office. In the office, they had about four or five men already. With him, they made up six. So they were given helmets,rifles and ammo belts and an order, "Go get that sniper." They were not born just yesterday. They all went out about halfway to the sniper and found themselves a crater from some explosion. They all laid back until the sniper left or stopped firing. They waited a little bit longer and then hustled back to the office where they turned in all the equipment given to them. Now off he went to Okinawa, his old stomping grounds, and his Okinawan girlfriend. He had a great time; He stayed at her home and went off to the movies and dinner. She took him places where only Okinawan lived and shopped. He was living the life.

One day, he looked out her window and saw this marine and his followers going down the street; he was doing karate moves. Apparently, he was a black belt in karate, and everyone was afraid of him. He remembers telling himself how stupid he looked. During that same night, he walked into a bar for a drink, and there was the karate marine. He made a beeline right to him and asked, "Who in the fuck are you?" he replied, "It's none of your business." He started circling around him, and everyone stood back. He then came at him, trying to kick him in the balls, but missed and ended up kicking his leg instead. At that moment, someone came from behind him and grabbed his arms and took him outside the bar. It was a big black fellow that took him outside. He tells him, "You don't want to mess with that guy because he'll kick your ass. He went on to say, everyone is afraid of him. He's a black belt in karate." The young man had taken karate himself when he was a kid. He knew how dangerous a black belt could be, so he thanked the guy for pulling him out of the bar. Then he went on his way to his girlfriend's house. It was time for him to go back to Vietnam, so he went to this hamburger place on top of the hill and ordered about a dozen hamburgers to take back with him. The young man was planning on giving someto some of the guys in his unit. Guess who was there at the same time? Yes, the black belt karate kid. The karate kid was sitting there, asking the young man, "Hey, you're the guy whose ass I kicked the other day." The young man replied, "You didn't kick my ass." So he turned and waited for his order of hamburgers. Then he started to walk up to the young man and said "Yah, you're the guy whose ass I kicked the other day," and he punched the young man right on his face. His punch didn't hurt at all. The young man was sitting on a stool. He ended up kicking the karate kid right at his balls. The young man gave the karate kid a big bear hug and lifted him off the ground. He ran with the karate kid across the restaurant, knocking down tables and chairs. The young man finally threw the karate kid on the ground, and he would alternate by hitting him in the balls and hitting his face. The young man will tell you, he was so focused on the karate kid that when the friend of the karate kid came over and hit the young man over the head with a steel chair, he didn't feel a thing. Some civilian Okinawans' came to help the young man and stopped the guy from hitting him with the chair again. He kept on hitting the karate kid until someone pulled him off. The restaurant owner called a cab for the young man. When the young man walked out, he could hear people saying "that little guy kicked the karate kid's ass"

The hamburgers never made it to Vietnam; they ate them right on the plane on their way to Vietnam. What can he say, they smelled great; the burgers were still hot, so they ate them on their way back. He didn't even know the guys. When they got to Vietnam, they all parted and went their different ways. He learned that A Battery 1/13 was located in Hill 10; at least the barracks and office were there. The battery was just a few clicks away. It looked like the Battery all arrived at the same time. Everyone was just settling in. He remembers when a Major Green pulled up to four of them just sitting on the ground, talking. Oh, he had on a brand-new uniform, shiny boots, a driver, and a clean jeep.

He jumped out of that jeep and told them to get at attention. He called them everything in the book. He told them they were unsquared away. Then the general asked them where he could find their captain. They told him he was just a couple of clicks down the road. The major said he was going down there to report them. They gave the general a warning by telling him the roads hadn't been cleared yet. He said, "The hell with the roads," and off he and the driver went. About five minutes later, the marines heard an explosion and saw a cloud of smoke coming from the direction of their battery. Needless to say, there was no more Major Green, but miraculously, the driver lived.

His buddy, Tony Zavala, somehow ended up on the same hill with him. Now Hill 10 was real close to Da Nang. They wanted to go into Da Nang real bad. So one night, this one guy stole a jeep. Now there were five marines all headed to Da Nang. There were three of them in the backseat and two in the front. The young man was pretending to be a corpsman; one guy was moaning and groaning. There were checkpoints all the way down the road, but at each checkpoint, they would go into their act, and the guards would open the gate for the five marines. Until they reached the last gate; they were stopped and the guards told them, they were going to call a chopper for him. They told the guy moaning and gowning, "You got to go with the chopper, or we'll all burn for this." The guy said, "Hell no, I'm not going. Someone else go." So they all decided to turn back. They were driving like a bat out of hell. All the gates were open for them except the last gate to get into Hill 10. They got pulled over by one of the guys on guard duty. He told them to pull over to the side because the officer of the guard wanted to talk to them. So as they pulled over to the side, they decided to haul ass into Hill 10; they could hear the men on guard duty yelling and stating, "Halt or we'll shoot!" They had the right to shoot them. They all thought they were about to die, fortunately, for them, The guards didn't shoot. They ended up parking the jeep down the hill. They all ran into their sleeping quarters and covered up like nothing had happened.

Well, you know marines: once their minds are made up, there is no stopping them. It wasn't long before they tried it again. This time, it was still daylight. He remembers they parked the jeep out on the street somewhere. They began their search for women. It wasn't long before the MPs caught them walking down the street; they were busted. Fortunately for them, a navy captain was coming out of this building right where they were being picked up. The captain asked, "What's going on here?" They explained to the captain that they had just come down from Khe Sanh, and they just wanted to see Da Nang. The captain convinced the MPs to place the marines in his care and that he would call one of his drivers to take them back to their unit. When the driver arrived, they all got into this little gray his bus, the driver, asked them where they wanted to go. They all responded, "Back to our jeep," so he drove them to their jeep, and they parted ways, but not before asking him for directions to some women. He pointed to a building nearby, and they drove over there, parked the jeep and went inside the building. There were plenty of women, so they had themselves a party. When they were all done, they took off to try to get back to their battery, but the road was closed. They pulled in to some unit because the gate was closed, and they weren't allowed to go through. So they were told they could spend the night with the unit. However, the first sergeant wanted to see them first thing in the morning. They all got up before sunup, got the jeep, and rolled it out of the compound. They all got in the jeep; they started it up and left. But once again, they were stopped at the beginning of the road to Hill 10 by the gatekeepers. They were told they had to wait until they finished clearing the road from booby traps and mines. The gatekeepers, said that during the night, the road belonged to the NVA and VC and that in the mornings we had to clear the roads, so the roads became ours again. Just then, a big deuce-and-a-half came right behind us. They were also getting on the guards; stating they needed to go, unload their load and be back before dark. So the guards finally gave up and said, "If you guys want to get killed, go right ahead." They let the marines go through, and they drove like a bat out of hell. For some reason, all the gates opened up and allowed them to go through. They went to park the jeep, and they all ran to our beds and began to play as though they were just waking up. There they were grinning from ear to ear, mission accomplished.

One day, they heard over the radio that a chopper was coming in with a boatload of captured enemy weapons. Since they were coming to their LZ, they walked over there to take a look. Well, it wasn't enemy weapons they brought over. It was seven dead marines on six stretchers; these men were cut down into pieces. They lined up the marines' torsos; on the stretcher. They had all the body parts. These poor guys didn't even die right. On the stretcher with all the body parts, on top of the heap, was a marine's head with his eyes still open. He looked as though he was looking toward home. The corpsmen were lifting up a torso by the collar of his uniform. This marine had no head; his arm was dangling by a nerve. The corpsmen were trying to empty out his pockets. As they searched his pockets, the arm fell off; it was a gruesome sight. These men were out on patrol and were ambushed in the worst way. The young man had a camera with him, but out of respect for the men, he didn't take their pictures. His friend, Tony was with him that day, so they both witnessed these atrocities. It was a terrible scene, one that would live with them forever.

One day, he was sent to deliver some classified material; the chopper took him to a few other hills first to pick up other men that were also delivering classified material. He remembers landing on this one hill, and there was this young second lieutenant with his hands full of loose documents. Well, the wind created from the chopper blades just blew away all his paperwork. His papers went flying all over the place. The chopper's blades were not helping him at all. He was given three days to deliver the classified material and return. All he took was his sidearm, his .45-mm pistol, with about five rounds in it, and he put the classified material inside his shirt; that way, he knew it was safe, and it wouldn't get lost like that second lieutenant's documents. He knew his way around Vietnam well, so he delivered the documents the very first day. Then the young man went over to the White Elephant, it was a navy base. He had himself a few drinks, went over and took a hot shower, and actually slept in a real bed. He'll tell you, those navy guys had it made; they worked hard, but they also got to play hard as well. It was time for him to head back. Wouldn't you know it? He got to the gatekeepers about ten minutes before they closed the road.

They decided he was not going on foot to his unit. The young man kept on arguing with the gatekeepers. He told them the gate was still open and he could easily make it to his unit on foot. The gatekeepers said, "Hey, the NVA and VC own the road now," but if you wanted to get yourself killed, go right ahead." He was humping it up the road; when he was about midway, when it started getting dark. Now, he tells himself, maybe he should have stayed back like the guards at the gate had advised. He pulled out his.45 sidearm and chambered a round. When he entered an area with a lot of bushes and a big boulder nearby, that's when he noticed a canoe coming down the stream of water. At first, all he could see was this old man wearing white silk clothing. Then the old man grabbed a hold of the front of the canoe and pulled it up on land right where the young man was located. The old man couldn't see the young man because of all the brush. Then out popped a young VC wearing black silk clothing carrying an AK-47. The young man had his gun pointing right at the young VC while he had his rifle pointing away from me. They were only twenty to thirty feet apart from each other. They looked directly into each other's eyes, and somehow they both came to the same conclusion that this was not a day to die. So the young VC turned away and proceeded on his way, and the young marine did the same. Shortly thereafter, a truck was coming by, driving like a bat out of hell. They stopped and asked the young marine if wanted a lift. He said, "Absolutely," so he got to ride in the back of the truck the rest to his unit. Then the truck moved on to another location.

He had no idea where they were. He felt lost, but he knew they were still close to Da Nang. He was getting short. He typed out a set of R

& R orders for himself and he even signed them. He did this because he heard they were no longer letting transients go out on liberty in Okinawa, and he wanted to see his girlfriend. Good thing he did because they were not letting any transients go out on liberty. So he was prepared. He put on civilian clothing, called a cab. He had the cabdriver drive him into town, which was right across the base. He flashed his R & R orders at the guards guarding the gate, and they let him right through, and off he went looking for his girlfriend.

He went to her, and he'll tell you, she was a little china doll. He almost brought her home with him. Yes, he got drunk and asked her to marry him. He is so glad she was sober and thinking straight because she said no, and besides, she was a street-working girl. She knew that it wouldn't have worked out, so they said their good-byes. Upon returning to base, he reported back to base; he learned he was put on a priority list for a flight home because he had was about five days left in the corps and he was still overseas. He flew back home from Okinawa and reported to Camp Pendleton for discharge. He learned that if you reported any problems with your health, they held you back for treatment. So he didn't tell them about the sweating he was having at night nor that he was afraid to put his head down on a pillow because of the nightmares he was having. He was supposed to be released on August 30, 1968; instead, they let him out on August 27, 1968. Since his sea-bag was destroyed in Vietnam, they gave him a complete set of uniforms and boots before they turned him loose. You see, he still had three more years of non-active or reserve duty to complete. The first thing he did when he got home was buy himself a new car. He bought himself a 1968 Chevy that was canary yellow, and he was living at home. His big regret is that he didn't share the money his mom saved for him while he was overseas. He guessed it was because he had big plans for himself. He wanted a new car. He wanted to find himself a good-paying job and get married with his beautiful girlfriend back in East Los Angeles.

Chapter 20:
New Chapter of Life

He asked her to marry him, but she said she couldn't right then and that he would have to wait. He went and applied for a job with Sears Company. He had to take a test. When the test was over, they offered him a job as a shipping-and-receiving clerk. You see where this is headed, a clerk again. He stayed there for about six months, then there was this Electric Motors Company doing some hiring, so he took the job; it paid ten cents more per hour. There he was putting wire into these electric motors. He got very good at it, and before he knew it, he was just cranking them out like hotcakes. The old-timers asked me to slow down—that he was making them look bad. But he ignored their requests, and before he knew it, his motors were no longer passing inspection. The work was being sent back to him for correction. As soon as he slowed down, his motors started passing inspection. It was a lesson he learned in life. Listen to your elders because they'll make you or break you in a heartbeat. From there he went over to Chrysler; they were just up the street from these guys. So, he applied for a job and was sent to an assembly line putting up dashboards and door strikers. Now this was a much better-paying job. It was hard work, but it paid well. Now he decided to move out. He went and got himself a small bachelor's apartment for fifty dollars per month. He started to go to school at night. He went to LA Trade Tech and took up color-TV servicing and math. The VA wasn't doing very good in processing claims back then. So here he was just one week left of school and with no money. He was eating bread and mayonnaise

The VA never processed his educational claim, so with only one week remaining, he dropped out from his classes. He was spending all his money on his girlfriend and other girls.

That's why he was broke. Well, his girl ended up dumping him. There was this guy she liked, but he was married. He used to work for Sears—yes, Sears—at the same place he did as a shipping-and-receiving clerk. In fact, they got together while he was in Vietnam. When she dumped him, he went crazy; then he knew how the lieutenant felt when he got his Dear John letter at Khe Sanh. During this time, he was self-medicating himself. He would mostly drink beer. He would drink about two or three beers every night, get up early in the morning, go to work, come home, and start drinking again. He had no idea what was wrong with him. Well, the union and Chrysler were not getting along. The union would call wildcat strikes all the time; one day, Chrysler just closed their doors, and they were all out of a job. He finally went over to this strange job. It was all gray; there was no color in it at all. Then he would see a flash of fire shoot across the walking path. He could make out men; it looked as though they were hooked up to black rubber hoses and chains. They were handling these funny-looking machines that would spit out fire. As he walked down the walkway, he remembers telling himself, this looks like hell. He wasn't sure if he should step through and into this porthole when all of a sudden, he sees red; it stood out like red stood out on the green leaves of Vietnam. A guy had cut himself, and blood was shooting out just like in 'Nam. Then someone grabbed him on the shoulder, and it turned out to be the foremen of that area. He told him, "Follow me." They walked up to where the guy was working; the foreman told the guy that was bleeding to go see the nurse. The foreman started to show him his job. He picked up on it very quickly, and then the foreman left him on his own. You see, the job he took was at Ford Motor Company in Pico Rivera, and the pay was even better than the his prior job. He started taking this girl out that lived right next door to his mom's house. She was a fine-looking woman. Her name was Kristen; he used to call her Kris. She was a bookworm and was very smart. They took a liking to each other, and before he knew it, she ended up pregnant. Well, they ended up getting married in Las Vegas. They drove home. He dropped her off at her home, and he went to his. By this time, his mother already had Sylvia, his only sister. Could you imagine that, six boys and one daughter? But right after Sylvia, came Robert. Now there were seven boys and one girl. Soon thereafter, he went and got them an apartment, fully furnished, and this was where he started his family. They had their first daughter and named her Elizabeth, and eleven months later, they had their second daughter whom was named Jamie. He was very strict with his daughters and became a pain in the ass to the wife. Even though he was mostly drinking beer, he started drinking at work, on the way home, and at home. He would finish two to three beers per night. He became self- destructive. He remembers wanting to kill himself. He just kept on getting these flashbacks. It was like looking at pictures; they would just pop into his head. He would see dead guys and guys with no heads. He would hear firefights and choppers. He would break out in a heavy sweat. Sometimes he would tremble at night.

One time, one night, he was driving home from a party. He was driving a brand new 1968 blue Camaro at the time. When he came up on this red light, he stopped the car, waiting for the light to turn green. And these two guys, a big one and a short one, they yelled out to him, "What time is it?" He tells them. Just when, the light turned green, and he started on his way when they threw this bottle of beer at him. They almost hit him on the head with that bottle. It hit the rim of the window and left a dent on his car. He stopped his car and assessed the situation: two against one, one guy was big bigger than he was. Anyway, should he have just keep going or go back? So he put his car in reverse. He stopped his car where they were and got out. Both of them started walking towards him. As soon as they were within arm's reach, he hit the big guy with his right fist, directly in his throat; the big guy toppled over, and the little guy began to run away.

The young man started beating on this guy with his elbows because in a previous fight the day earlier, he hadb broken his thumb. So here he was just beating this guy. He pushed back the guys front teeth, busted his nose, and broke his jaw. The young man had blood all over himself. He tells the guy if he moved that he was going to kill him. So he didn't move. He finally stopped beating on him. They both stood up, the young man asked. "Why did you throw that bottle at me?" He said he was sorry. And then the bg guy asked if he could pull his teeth back out? The young man said he would, but if the big guy bit him, he was a dead man for sure. He reached in there with his finger and grabbed his teeth, and they all popped out and fell in place. Just then, the police arrived, and the young man explained what had happened. They asked him if he had hit the guy with a tire iron. The young man responded with a no. The cops could see where the bottle hit his car, made a dent on it, plus spraying the car with beer. The cops cuffed the big guy and took him in. They just sent the young man on his way. The young man was unafraid and highly trained to kill. He had no business being out on the streets in his condition. But he didn't know what was wrong with him.

Just about a day before, he met this guy; he was driving right down Whittier Boulevard when a car backed right into the front end of his vehicle. They both got out to see the damage, but before the young man knew it, the other driver jumped back into his vehicle and sped away down Whittier Boulevard. He didn't give the young man a chance to see what kind of damage had been done. He had his good friend Frank with him in the car. Frank was married to Katie his childhood girlfriend. She was a beautiful looking woman. The young man got right behind the other driver, so when the light turned red, they had to stop. The young man got out of his car and walked over to the other car. The window was rolled down, so the young man was asking the other driver to pull over. But when the light turned green, the other guy started taking off. The young man immediately jumped through the open window. He had the driver laid out on the front seat of his car while the young man was steering the vehicle with his left hand and beating on the other driver with his right. He was halfway out of the vehicle, and the guy he was beating, was stepping on the gas. They were going around cars in the middle of Whittier Boulevard. The other driver had a carload of guys. So, they all eventually told the other driver to step on the breaks, which he did. Then they all started hitting the young man and finally threw the young man out of the vehicle. There he was in the middle of the street wondering how they missed all those cars, when his friend Frank pulled up with the young man's vehicle.

Unknown to him, another car also gave chase and made the other driver crash on the sidewalk; his car was being held up by a parking meter. When his friend and he drove off, they didn't expect to see the other guy anymore. But as they were driving, there he was with the front end of his car up in the air; all his friends had bailed out and left him alone. There he was burning rubber, trying to get himself unstuck from the parking meter. The young man arrived and pulled up on the corner of the street. The young man just had to walk a few stores down to reach him. He pulled the other driver right out of his car and gave him a right cross. Unfortunately for the young man, he broke his thumb when he hit the other guy. But that's all he needed because thother driver dropped down to his knees. He couldn't fight anymore; he was in a daze. The police showed up. The young man explained what had happened. They walked over to his car to check out the damage. There was a little chip from the front grill that was broken off. He explained to the officer that the other guy didn't give him a chance to look at the damage. So the officer just sent the young man on his way. The young man had other fights. He won them all. He never started them, but he sure as heck finished them. He was just in a bad way. He needed treatment but just didn't know it.

Unfortunately, he treated the kids like if they were in boot camp. He treated the wife, Kris, the same way. He is not making any excuses, but he needed to treat them better. The wife started looking for a way out of her situation. She did; He eventually threw her out. She never came back. He did go to her work site and asked her to come home; this took place about two weeks after he threw her out. He gave her an ultimatum: either she came home with him right there and then or he was going to file for a divorce. She chose to stay there, so off he went to see a lawyer. They ended up in court. Everything went so fast; his head was spinning, and before he knew it, they were divorced for irreconcilable differences. He got to keep the kids, which was primarily his main objective. To him nothing else mattered.v

Chapter 21:
Newfound Love

About three or four years later, she moved back into the neighborhood; she and her new husband bought a house just three blocks away from them. They were not hateful toward each other. In fact, they remained friends. In fact, her new husband was also a marine, so they hit it off good. The young man went fishing with him on his boat, freshwater fishing; we had a great time. In a divorce, it's best to remain friends for the kids' sake, and till this day, they still remain friends. Back at his job at Ford Motor Company, some of the guys would make fun of him because he was going to school at night. They would ask him why he must go through all that trouble because he was going to end up retiring from Ford Motor Company. All that schooling was just going to waste. Well, in spite of those remarks, he still continued to go to school. he learned to enjoy school, so he wouldn't have it any other way. Well, he went to Cerritos College for about eight to ten years, mostly going part-time. He had amassed a total of one hundred and ten units. He was able to obtain an AA degree on social welfare. He got it just in time to because they all learned they were closing the plant. The plant closed just before Christmas.

He applied at ten different locations. They all called him to report for work. He had his three top picks: the VA, US Post Office, and a counseling job for his city. He remembers pacing all night, trying to decide which one would be best for him. He finally decided to go to work with the VA. The VA was where he saw his dad die at the VA medical center. Little did he know that someday he would be working for the VA Regional Office, which was right across the freeway from the hospital. He was a veterans' benefits councilor. So he went from a GS-5 to a GS-7, then he went to a GS-9. He told his new wife he wasn't satisfied, so he put in for a field position, making him a GS-10. He had his own office in downtown Los Angeles. He was there for about four to five years. He had a lot of opportunities to take out some ladies, but he made it a rule to never mix business with pleasure, he sure lost some fine opportunities. Working out in the field gave him the opportunities to work at different locations. He remembers working at the VA Long Beach; there was this girl that kept asking questions about him. She boldly came out, asking for a date to go out whale watching. She was very smart and beautiful. Well, she caught him on the rebound, and he accepted. They went out with one of her friends. They had lots of fun and enjoyed each other's company.

This was like in January 1980. They became good friends. However, her family didn't approve of him. They didn't like the fact that he was brown, previously married, and had children to raise. So he went to her home and talked to her father. He told her father, he was taking his daughter and that they were going to live together. He was fit-to-be-tied. He thinks her father wanted to hit him right in the face, but he held back. This girl's name is Ellenmarie Brockbank; she was very smart and as pretty as could be. She had big hazel eyes and was beautiful, just natural. She didn't have to wear all that makeup to look good. She was a pretty and fine looking woman as well. She was also well-organized person. Well, his cousin Bill Jimenez helped him move her out; they used his van to load all her belongings. They must have made two or three trips; she was finally all moved in. They were basically working and coming home. She agreed they wouldn't have any children, so they were both working for the VA at that time. She found an opening at the IRS (Internal Revenue Service), so she put in for the job and got it. She was doing great, got herself promoted, and at times, they would put her in as acting supervisor. They did a feature on her in the Los Angeles

Times. They were doing great, and he started doing some thinking. Here he was with two daughters to raise, and Ellenmarie would never know what it would be like to have any children of her own. He asked her, "Look, we only have a short window to have some children. Let's take advantage of this opportunity." So they did, and here came Shannon Lynne. She was a good-looking little baby, and two years later, here came Annamarie Dolores. We were blessed with two healthy and good-looking little girls. Ellen had quit her job when she had Shannon, so she was ready when she had Annamarie. As for him, it was time to move on. He tells the wife he was going for the next position. He got to be a supervisor, GS-11; now he had his own unit. He was really lucky to have the best unit in the VA; the people he had were just fantastic. They knew what had to be done, and they did it. They really made him look good. So when the position opened up for admin chief, he put in for the job. Would you believe it? Here he was—went full circle. The marine corps must have known he was an Admin person because here he was being promoted to Admin Chief or the Administrative Officer of the VA Regional Office in West Los Angeles, and he loved it. It was the best job in the whole VA system. While in this position, he was awarded five Hammer Awards for making the government more efficient, thus saving the VA over nine hundred thousand on one project alone. He was recognized for his contributions while serving on the board of Directors to CASU, an organization that helped government agencies make large purchases without a lot of red tape. He was also recognized for his contributions to the Federal Executive Board for the Greater Los Angeles area. He was now a GM-13, step eight. The young man would like to remind you that your mind is a very powerful muscle. Exercise it, meditate, concentrate and focus on a single point, you will be pleasantly surprised as to how much power and strength you'll be able to harness. The things that you'll be able to accomplish are limitless. Don't be afraid to open portholes, they are all around us. You could be a subject matter expert on any subject of interest. Of the Marines he served with, they were the best people he ever served. Most were sons and daughters from what most call the Greatest Generation.

They came from Main, Kentucky, New Mexico, California, and across America. They were the toughest and the best people America had. For those who died; they didn't die for flag, country, or apple pie. They died for each other for they became brothers.

Family Ancestry

Family ancestry is a fascinating science of study. One's genetic makeup, your DNA, contains billions of imprinted materials that makes you so unique and special from everyone else in the universe. The information imprinted in everyone determines which direction each individual leans.

DNA is not the only influence in one's individual makeup. Like a pine tree, a pine tree is not just a pine tree. There is a variety of elements that will have a tremendous influence on the pine trees growth. Examples are: its environment, is the ground nutritionally enough for good growth? Will deer eat vital branches? Will the surrounding trees impact its growth? Will fire have a negative impact on its growth? Or will some or all make this pine tree stronger and more resilient than ever?

Some theorists believe that the further back in time one goes, one will eventually find themselves in the future. Just look at the ancient Mayans and Aztecs. What you will find to be so incredible is that they had advance knowledge of mathematics, engineering, astrology, agriculture, writing, and religious beliefs.

It is proven that cultures around the world had similar knowledge all around the world. There are ancient structures, cartouches and writings that are screaming to impart advance knowledge to anyone capable of interpreting the evidence left behind.

Below are photos of the young man's grandparents and family members caring the DNA from their ancient ancestors. What an amazing life.

The Young man's Family (Grandparents)

"FamilyMembers"

Khe Sanh